AF581711

New Advances of Cavitation Instabilities

New Advances of Cavitation Instabilities

Editor

Florent Ravelet

MDPI • Basel • Beijing • Wuhan • Barcelona • Belgrade • Manchester • Tokyo • Cluj • Tianjin

Editor
Florent Ravelet
LIFSE
Arts et Metiers Institute of
Technology
Paris
France

Editorial Office
MDPI
St. Alban-Anlage 66
4052 Basel, Switzerland

This is a reprint of articles from the Special Issue published online in the open access journal *Applied Sciences* (ISSN 2076-3417) (available at: www.mdpi.com/journal/applsci/special_issues/cavitation_instabilities).

For citation purposes, cite each article independently as indicated on the article page online and as indicated below:

LastName, A.A.; LastName, B.B.; LastName, C.C. Article Title. *Journal Name* **Year**, *Volume Number*, Page Range.

ISBN 978-3-0365-1546-5 (Hbk)
ISBN 978-3-0365-1545-8 (PDF)

Contents

Editorial

Editorial for Special Issue: New Advances of Cavitation Instabilities

Florent Ravelet 

Arts et Métiers Institute of Technology, CNAM, LIFSE, HESAM University, 75013 Paris, France; florent.ravelet@ensam.eu

Abstract: This editorial presents the main articles published in the Special Issue: New Advances of Cavitation Instabilities.

Keywords: water jet cavitation; cavitation peening; hydrodynamic cavitation; compressible two-phase flow; Kelvin-Helmholtz instability; slamming; vortex rope; bulb turbine; hydrofoil; Francis turbine

Citation: Ravelet, F. Editorial for Special Issue: New Advances of Cavitation Instabilities. *Appl. Sci.* **2021**, *11*, 5313. https://doi.org/10.3390/app11125313

Received: 1 June 2021
Accepted: 4 June 2021
Published: 8 June 2021

Cavitation refers to the formation of vapor cavities in a liquid when the local pressure becomes lower that the saturation pressure [1]. In many hydraulic applications, cavitation is considered as a non-desirable phenomenon, as far as it may cause performance degradation, vibration problems, enhance broad-band noise-emission and eventually trigger erosion.

In this Special Issue, the deliberate use of cavitating jets for peening treatment of materials has been considered by Yoshimura et al. [2] and by Soyama [3]. This very interesting review shows how the aggressive intensity of a cavitating jet varies non-monotonically with the cavitation number in accordance with the variation of mixture sound speed for different regimes and that there exists an optimum.

These findings illustrate the importance of instabilities in cavitating flows. The cavitation is usually associated with flows with very high Reynolds numbers. This formally unsteady flow type is very sensitive to disturbances. In the ordinary case of a 2D profile, it is rightly recognized that for sufficiently low cavitation numbers, periodic or quasi-periodic cavity shedding arises. On more complex problems, like cavitation in inducers, several other periodical behaviors can emerge. Firstly, the local intrinsic flow instabilities will depend on the region that presents hydrodynamic cavitation: tip-leakage vortices, back-flow vortices, or blades suction surface. Secondly, interaction between adjacent blades can lead to rotating instability, with cells of various sizes that propagate from blade to blade. Finally, system instabilities are also observed, because of the blockage linked to the volume variation of the pockets, or to a possible positive slope of the inducer characteristics linked to a change in the angle of attack on the blades. A more proper understanding of these instabilities is of crucial interest, especially in the field of turbomachinery, still motivating applied and fundamental research on cavitation instabilities.

In this Special Issue, Pipp et al. [4] and Ravelet et al. [5] present new findings in cloud shedding mechanisms arising in a 2D venturi geometry, respectively at very small scale (micro channel) and at at very low Reynolds numbers, where the development of the Kelvin-Helmholtz instability is evidenced. Viitanen et al. [6] studied numerically the cavitation dynamics on a standard hydrofoil with a compressible approach and a delayed-detached eddy simulation and show evidence of pressure wave fronts associated with a cavity cloud collapse. Pancirolli et al. [7] use a smooth particle hydrodynamics (SPH) model to study the cavity formation during the impact of a 2D wedge on a free surface and show that there is a threshold for the ratio between horizontal and vertical velocities above which a cavity is created, greatly impacting the stability of the body during slamming.

On a more applied point of view, Podnar et al. [8] improved the shape of an hydrofoil in order to reduce the cavitation development in a bulb turbine. Decaix et al. [9] conducted a comprehensive investigation on the importance of the time step tuning to achieve a good prediction on the cavitating behavior of a Francis Turbine. Finally, Li et al. [10] numerically explored the shape optimization of a globe valve in order to suppress cavitation inside of it.

Conflicts of Interest: The author declares no conflict of interest.

References

1. Brennen, C. *Cavitation and Bubble Dynamics*; Cambridge University Press: Cambridge, UK, 2013. [CrossRef]
2. Yoshimura, T.; Iwamoto, M.; Ogi, T.; Kato, F.; Ijiri, M.; Kikuchi, S. Peening Natural Aging of Aluminum Alloy by Ultra-High-Temperature and High-Pressure Cavitation. *Appl. Sci.* **2021**, *11*, 2894. [CrossRef]
3. Soyama, H. Cavitating Jet: A Review. *Appl. Sci.* **2020**, *10*, 7280. [CrossRef]
4. Pipp, P.; Hočevar, M.; Dular, M. Numerical Insight into the Kelvin-Helmholtz Instability Appearance in Cavitating Flow. *Appl. Sci.* **2021**, *11*, 2644. [CrossRef]
5. Ravelet, F.; Danlos, A.; Bakir, F.; Croci, K.; Khelladi, S.; Sarraf, C. Development of Attached Cavitation at Very Low Reynolds Numbers from Partial to Super-Cavitation. *Appl. Sci.* **2020**, *10*, 7350. [CrossRef]
6. Viitanen, V.; Sipilä, T.; Sánchez-Caja, A.; Siikonen, T. Compressible Two-Phase Viscous Flow Investigations of Cavitation Dynamics for the ITTC Standard Cavitator. *Appl. Sci.* **2020**, *10*, 6985. [CrossRef]
7. Panciroli, R.; Minak, G. Cavity Formation during Asymmetric Water Entry of Rigid Bodies. *Appl. Sci.* **2021**, *11*, 2029. [CrossRef]
8. Podnar, A.; Hočevar, M.; Novak, L.; Dular, M. Analysis of Bulb Turbine Hydrofoil Cavitation. *Appl. Sci.* **2021**, *11*, 2639. [CrossRef]
9. Decaix, J.; Müller, A.; Favrel, A.; Avellan, F.; Münch-Alligné, C. Investigation of the Time Resolution Set Up Used to Compute the Full Load Vortex Rope in a Francis Turbine. *Appl. Sci.* **2021**, *11*, 1168. [CrossRef]
10. Li, J.; Gao, Z.; Wu, H.; Jin, Z. Numerical Investigation of Methodologies for Cavitation Suppression Inside Globe Valves. *Appl. Sci.* **2020**, *10*, 5541. [CrossRef]

Article

Peening Natural Aging of Aluminum Alloy by Ultra-High-Temperature and High-Pressure Cavitation

Toshihiko Yoshimura [1,*], Masayoshi Iwamoto [1], Takayuki Ogi [1], Fumihiro Kato [1], Masataka Ijiri [2] and Shoichi Kikuchi [3]

1 Department of Mechanical Engineering, Sanyo-Onoda City University, 1-1-1 Daigaku-Dori, Sanyo-Onoda, Yamaguchi 756-0884, Japan; f117008@ed.socu.ac.jp (M.I.); f120604@ed.socu.ac.jp (T.O.); f120605@ed.socu.ac.jp (F.K.)
2 Department of Advanced Machinery Engineering, Tokyo Denki University, 5 Senju-Asahi-Cho, Adachi-Ku, Tokyo 120-8551, Japan; ijiri@mail.dendai.ac.jp
3 Department of Mechanical Engineering, Shizuoka University, 3-5-1 Johoku Naka-ku, Hamamatsu, Shizuoka 432-8561, Japan; kikuchi.shoichi@shizuoka.ac.jp
* Correspondence: yoshimura-t@rs.socu.ac.jp; Tel.: +81-836-88-4562

Featured Application: Mechanical structural parts such as automobile parts made of aluminum alloy that have long-term stable compressive residual stress and require an extremely hard surface.

Abstract: The peening solution treatment was performed on AC4CH aluminum alloy by ultra-high-temperature and high-pressure cavitation (UTPC) processing, and the peening natural aging was examined. Furthermore, peening artificial aging treatment by low-temperature and low-pressure cavitation (LTPC) was performed, and the time course of peening natural aging and peening artificial aging were compared and investigated. It was found that when the AC4CH alloy is processed for an appropriate time by UTPC processing, compressive residual stress is applied and natural aging occurs. In addition, the UTPC processing conditions for peening natural aging treatment with high compressive residual stress and surface hardness were clarified. After peening artificial aging by LTPC processing, the compressive residual stress decreases slightly over time, but the compression residual stress becomes constant by peening natural aging through UTPC treatment. In contrast, it was found that neither natural nor artificial peening natural aging occurs after processing for a short time.

Keywords: multifunction cavitation; water jet cavitation; ultrasonic cavitation; high-temperature high-pressure cavitation; peening natural aging; low-temperature low-pressure cavitation; peening aging

Citation: Yoshimura, T.; Iwamoto, M.; Ogi, T.; Kato, F.; Ijiri, M.; Kikuchi, S. Peening Natural Aging of Aluminum Alloy by Ultra-High-Temperature and High-Pressure Cavitation. *Appl. Sci.* **2021**, *11*, 2894. https://doi.org/10.3390/app11072894

Academic Editors: Marwan Al-Haik and Theodore Matikas

Received: 6 February 2021
Accepted: 22 March 2021
Published: 24 March 2021

Publisher's Note: MDPI stays neutral with regard to jurisdictional claims in published maps and institutional affiliations.

1. Introduction

The mechanism of high-pressure water jet action on materials has been well understood for decades [1]. Shot peening [2,3] and fine particle peening [4] are techniques for mechanically striking surfaces with particles to impart compressive residual stress. Another processing technology modifies surfaces with an abrasive water jet formed by adding an abrasive to water pressurized to about 300 MPa [5,6].

When water jet cavitation (WJC) is irradiated with ultrasonic waves to increase the temperature and pressure, the cavitation has both mechanical and electrochemical affects. This is called multifunction cavitation [7,8] (MFC, PCT Pub. No. W02016/136656, US Registered Patent [9]). WJC has a bubble radius of about 100 μm, and the impact pressure at the time of bubble collapse is about 1000 MPa [10]. In contrast, in the ultrasonic cavitation (UC) used for cleaning, the bubble nuclei originally existing in the water with a radius of several micrometers are irradiated with ultrasonic waves, and the collapse pressure is small, but the temperature at the time of bubble shrinkage is several thousand kelvins [11–13]. MFC bubbles have a large radius of about 100 μm, a collapse temperature of several thousand kelvins, and a high collapse pressure of about 1000 MPa. Such bubbles are a feature of WJC and UC. The distribution and unsteady flow of WJC bubble clouds that are changed by sound pressure from incident ultrasonic waves have been clarified analytically and experimentally [14]. Previous studies have developed ultra-high-temperature and high-pressure cavitation (UTPC) obtained by attaching a swirl flow nozzle (SFN) used for water jet peening in air to further increase the bubble size, number of bubbles, temperature, and pressure of MFC [15]. In addition, a photon counting head was used to measure the photons (sonoluminescence) emitted from UTPC, MFC, UC, SFN-WJC, and WJC. The UTPC has the highest emission intensity, number of photons, and bubble interior temperature upon collapse. It was clarified that there is a correlation between the sonoluminescence results and the surface modification results of low-alloy steel [9]. Using these high-temperature and high-pressure bubbles, surface modifications of various metals and ceramics have been carried out.

The foremost use of aluminium casting alloys is in the automotive industry (60% of the tonnage world-wide). These are mainly engine components elaborated from alloys with silicon, most of which also contain at least 3% copper. The main applications outside the automotive industry are in mechanical construction, electrical engineering, transport, household electrical appliances and ironmongery [16]. Studies on the processing and natural aging of aluminum alloys have been also reported. The influence of the stirring pin and pressing tool shoulder on the microstructural softening during friction-stir processing and subsequent natural aging behavior was investigated for a 6061-T6 aluminum alloy [17]. Severe plastic deformation (SPD) can be used to generate ultra-fine-grained microstructures and thus to increase the strength of many materials. Unfortunately, high-strength aluminum alloys are generally hard to deform, which imposes severe limits on the feasibility of conventional SPD methods. The low-temperature equal-channel angular pressing method was used to deform an AA7075 alloy, and the influence of natural aging of a solid solution heat treated AA7075 alloy (prior to SPD) on the microstructure and the mechanical properties after equal-channel angular pressing was studied [18].

In this study, the outermost surface of a material was solution-treated using UTPC processing to raise the temperature to at least 500 °C higher than the equilibrium phase diagram of the aluminum alloy. Then, the elements added to the aluminum alloy formed a supersaturated solid solution and became distributed uniformly. After that, the possibility of T4 treatment, which is a natural aging process that gradually precipitates a Guinier–Preston (GP) zone at room temperature, was examined from the viewpoints of hardness, structural change, corrosion resistance, surface shape, and changes in compressive residual stress over time.

2. Materials and Methods

The specimens were Al-Mg-Si alloy (AC4CH), and its chemical composition is shown in Table 1. Rolled-plate tempered material of 30 mm × 30 mm × 5 mm was cut from the rolled state and used as specimens. Therefore, they were in a state where work hardening remained.

Table 1. Chemical composition of Al-Mg-Si alloy (AC4CH) (wt%).

Si	Mg	Fe	Ni	Ti	Al
6.80	0.31	0.10	0.01	0.01	Bal.

UTPC processing aimed at natural aging and low-temperature and low-pressure cavitation (LTPC) processing for comparison were performed on the rolled tempered material. Figure 1 is a schematic diagram of the equipment that performed UTPC and LTPC processing. High-pressure water is injected from a high-pressure pump and generates WJC, which is irradiated with ultrasonic waves from an ultrasonic transducer. The isothermal expansion and adiabatic compression [19,20] of WJC continues when the sound pressure around WJC remains below the break threshold [21] due to changes in the sound pressure of longitudinal waves. This produces high-temperature and high-pressure MFC. Furthermore, when a swirling nozzle is attached to the jet nozzle, the pressure of the jet center is reduced as the swirling flow is generated, the size of WJC increases, and the number of bubbles of WJC also increases due to the decrease of the cavitation number. When this WJC is irradiated with ultrasonic waves, UTPC is generated [15].

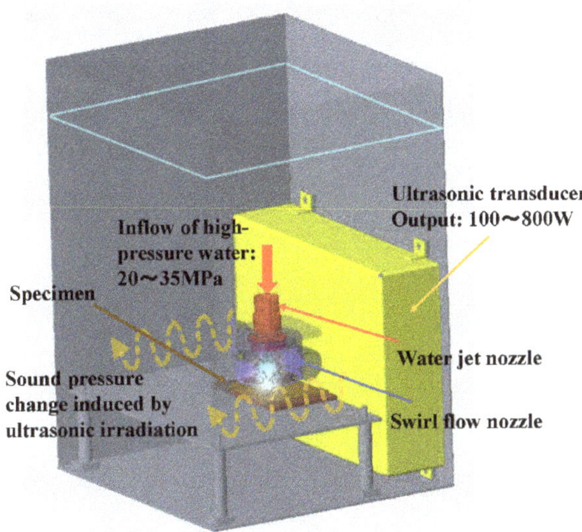

Figure 1. Schematic diagram of processing equipment for ultra-high-temperature and high-pressure cavitation (UTPC) and low-temperature and low-pressure cavitation (LTPC).

As the study on the cavitation number, there is a report that the formation of cloud cavitation vortex is controlled by the cavitation number in a narrow region similar to the swirling nozzle of this study [22]. According to the calculation of a single bubble using the Rayleigh-Plesset equation, for example, when a sound pressure of 130 kPa due to ultrasonic waves is applied to a WJC with a radius of 100 μm, the bubble expands to the bubble radius of 200 μm and shrinks to 3 μm in a time of 41 μs. The pressure reaches 3700 MPa and the temperature becomes 2.37×10^5 K inside the bubble. However, in fact, at the final stage of contraction, the increase in the bubble temperature is suppressed by chemical reaction heat and heat conduction accompanying the thermal decomposition of steam. In conventional measurement data of sonoluminescence, it is considered that the maximum bubble temperature is 100,000 K. The dynamics of single-bubble and multi-bubble was experimentally investigated using sonoluminescence [23]. It is known that the pressure and temperature achieved inside a bubble during multi-bubble sonoluminescence (MBSL) are much lower than those during single-bubble sonoluminescence (SBSL). In addition, when MFC or UTPC bubbles reach the minimum volume, they flatten in the direction parallel to the material surface, and the bubble part farthest from the surface becomes a piercing form, forming a high-speed jet (microjet) that pierces the material surface. However, the instability in microjet and two bubble interaction were reported [24]. Further studies are needed on the relationship between interactions in multi-bubble and microjet.

The pressure inside a bubble during collapse is mainly determined by the pump discharge pressure, and the temperature inside the bubble is mainly controlled by the ultrasonic output. The UTPC conditions were pump discharge pressure of 35 MPa, the standoff distance of 65 mm between the sample and the water jet nozzle, ultrasonic output of 800 W, and ultrasonic frequency of 28 kHz, whereas the LTPC conditions were pump discharge pressure of 20 MPa, ultrasonic output of 100 W, and ultrasonic frequency of 28 kHz. Here, the ultrasonic mode was selected as the dual mode because it had the highest peening effect in previous studies. The LTPC condition is the so-called peening aging [25] condition, in which the highest compressive residual stress more than −170 MPa can be applied to the AC4CH aluminum alloy. The sound pressure of ultrasonic waves controls the temperature of MFC bubbles. In this study, the sound pressure of ultrasonic output 800W that irradiates the water jet cavitation is 15 mV under the UTPC condition. On the other hand, the sound pressure of the ultrasonic output 100 W is 5 mV under the LTPC condition. These are the values of relative sound pressure measured by the sound pressure sensor (Portable Sonic Monitor: HUS-3, HONDA ELECTRONICS CO., LTD.). The processing time was changed to 2 min, 5 min, 10 min, 20 min, and 30 min to increase the outermost surface temperature to the solution heat temperature and clarify the time required for the solution treatment. Furthermore, to investigate the effect of processing for a longer time, UTPC for 60 min was also performed.

As a method for evaluating the peening natural aging by cavitation, Vickers micro-hardness measurement (Mitutoyo Corp., HM-210B system), observation and analysis by scanning electron microscope–energy dispersive X-ray spectroscopy (SEM-EDS) performed by a field emission SEM (Hitachi, S-4800), surface roughness measurement, and surface potential (corrosion potential) measurement with a Kelvin force probe microscope (KFM) (Hitachi, AFM5000II) were carried out. In the hardness measurement, the specific locations in the peening region to which the compressive residual stress was imparted were determined, and the change in hardness over time was examined. In addition, 10 points were measured at each processing time under each condition, and the average of 8 points with the maximum and minimum values removed was calculated. In addition, the compressive residual stress applied to the surface was measured by Cr Kα X-ray diffraction (MSF–3 M, Rigaku Corp.) to evaluate the change over time.

Figure 2 shows the principle of cavitation natural aging. As previously discussed, UTPC (micro-forging) is suitable for surface modification of refractory metals, and LTPC is suitable for low-melting-point metals, such as aluminum alloys. When LTPC is applied to an aluminum alloy, the microjet has few hot spots [11] and the temperature inside the bubbles is low, which is appropriate for artificial aging, so it is possible to impart compressive residual stress and artificial aging simultaneously. However, when a low-melting-point metal is processed by UTPC with many hot spots in the microjet, the additive elements of Mg and Si become a supersaturated solid solution at room temperature, but the outermost surface is overheated above the solution temperature. Therefore, Mg and Si are mixed uniformly. After the UTPC stops, a GP zone gradually precipitates due to natural aging, and the compressive residual stress formed by the elastic restraint of the surroundings creates new dislocations, which are entangled with the precipitates and hardened by the Orowan mechanism.

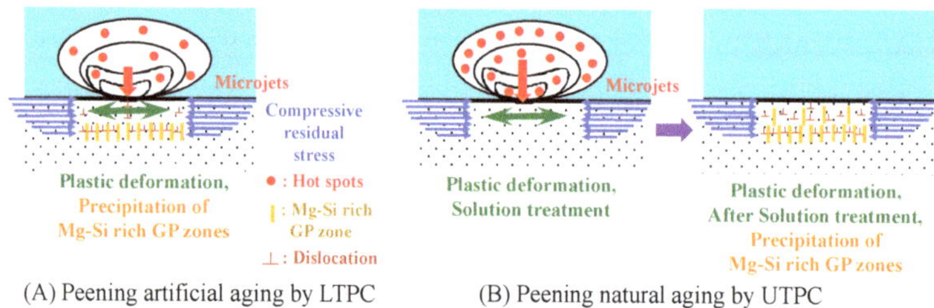

Figure 2. Schematic diagram of peening artificial aging by LTPC and peening natural aging by UTPC.

3. Results and Discussion

Figure 3 shows photographs of the LTPC processing surface jetted at a fixed point. When fixed-point injection is performed with using the SFN, an erosion region is formed at the injection center, and a peening region is formed around the transition region of the erosion region. In particular, the distribution of the peening region becomes pronounced after processing of 5 min or more. Figure 4 shows the photographs of UTPC processing surface jetted at a fixed point. Compared to LTPC, the peening area is expanded in the peripheral direction. Although the area of the perforated disc-shaped peening region of LTPC is small, the compressive residual stress is higher than that of UTPC, which has a large peening region, due to the peening artificial aging.

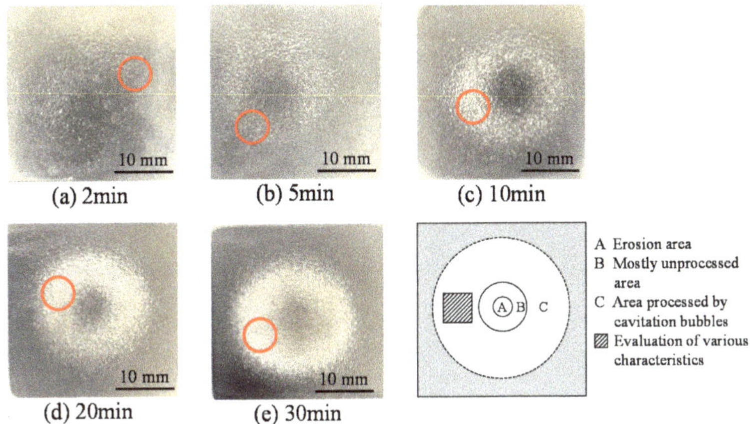

Figure 3. Surface photographs after LPTC processing (alloy: AC4CH, jet: 20 MPa, UC: 100 W).

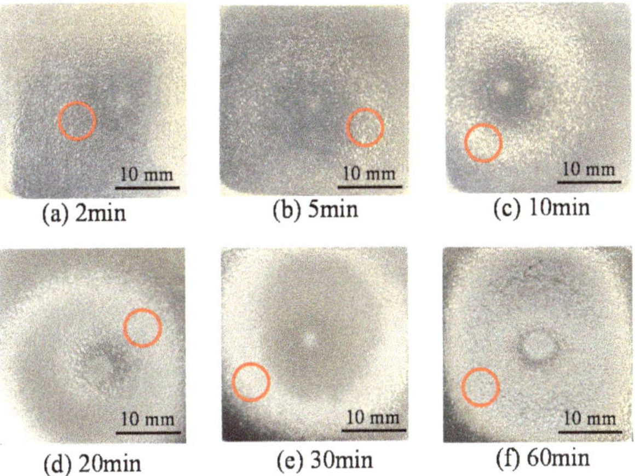

Figure 4. Surface photographs after UPTC processing (alloy: AC4CH, jet: 35 MPa, UC: 800 W). (○: roughness and hardness measurement area).

Figure 5 shows the changes of surface roughness with processing time in the peening region after UTPC and LTPC processing. Arithmetic mean roughness Ra increases with processing time, but there is no significant difference in surface roughness between UTPC and LTPC up to 20 min. However, from the KFM observation results, the number of peening marks is larger in LTPC than in UTPC, and the surface of LTPC is relatively smooth compared to that of UTPC, but the UTPC surface has many fine irregularities.

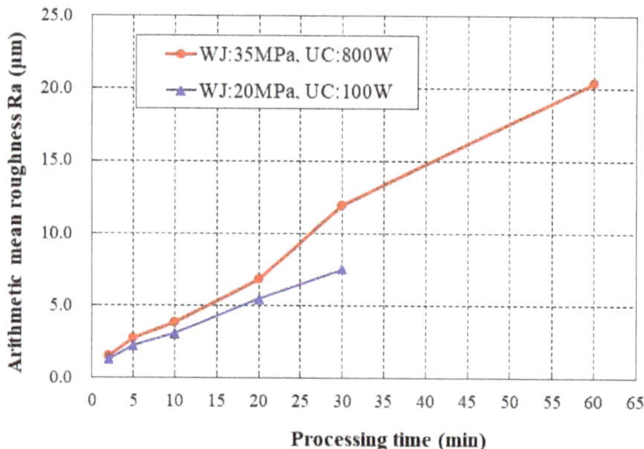

Figure 5. Change of surface roughness with processing time of LPTC and UTPC (LTPC: jet 20 MPa, UC 100 W; UTPC: jet 35 MPa, UC 800 W) (Evaluation length 5 mm)

The Vickers microhardness of the untreated material was 85.3 HV. Figure 6 shows the time course of hardness after UTPC processing at various times. No increase in hardness was observed after 2 min or 5 min of processing, but an increase in hardness was observed after processing for 10 min or more. The hardness of the material treated for 20 min increased the most. After 20 min of UTPC processing, the hardness reached its maximum value 21 days after processing, and then the hardness decreased and increased again.

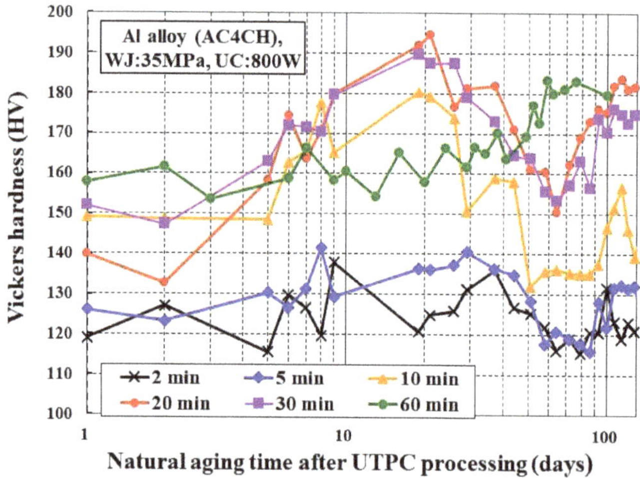

Figure 6. Relationship between hardness and elapsed time after UPTC processing (WJ: 35 MPa, UC: 800 W) (Average of standard deviation: 2 min 5.9 HV, 5 min 7.5 HV, 10 min 9.9 HV, 20 min 9.5 HV, 30 min 11.2 HV, 60 min 6.9 HV).

The reason the hardness increased after 10 min or more of UTPC processing is as follows. The outermost surface is solution treated by UTPC processing, as shown in Figure 2, and the supersaturated solid solution elements Mg and Si are perfectly mixed in aluminum at the solution temperature. After that, when cavitation stops and the temperature decreases to room temperature, the surface becomes a supersaturated solid solution, and Mg and Si gradually precipitate as GP zones. The surface is plastically deformed by the microjet when cavitation bubbles collapse, and compressive residual stress is imparted. This compressive stress causes dislocations on the surface, and the dislocations are trapped in the precipitates and hardened by the Orowan mechanism. The reason the hardness of material processed for 20 min and 30 min decreases at first and then increases again is thought to be a result of natural aging as the structure changes from GP I phase to GP II phase and θ′ phase. It is considered that the compressive residual stress generated causes new dislocations that are related to the pinning effect on the precipitate.

In contrast, it can be recognized that after 60 min of processing, unlike the material processed for 20 min, hardening does not occur immediately after processing. The hardness immediately after processing was about 160 HV, which was about the same as that for the LTPC 20-min processing and LTPC 30-min processing shown in Figure 7. However, after 60 days, the 60-min processed surface was hardened to 183 HV.

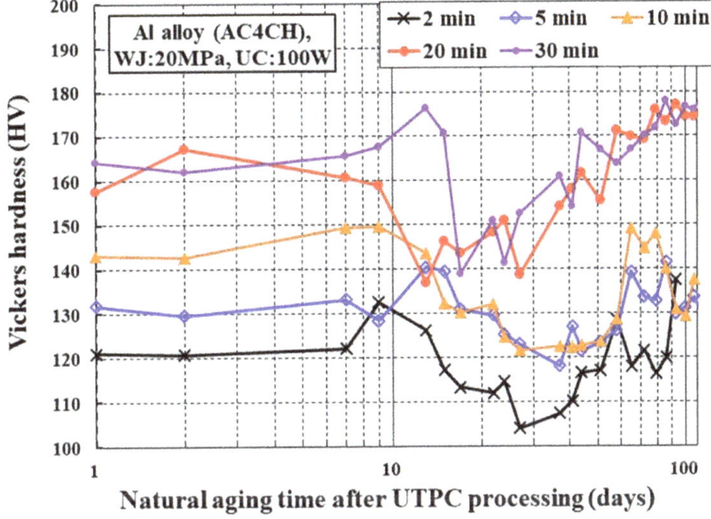

Figure 7. Relationship between hardness and elapsed time after LTPC processing. (WJ: 20 MPa, UC: 100 W) (Average of standard deviation: 2 min 7.7 HV, 5 min 6.6 HV, 10 min 8.5 HV, 20 min 8.4 HV, 30 min HV, 60 min 8.4 HV).

Al-0.62% Mg-0.39% Si alloy was quenched in water after solution treatment and naturally aged. The hardness increased by 12 HV on 1 day after solution treatment and quenching, and an increase by 22 HV on day 11 has been reported [26]. Because the chemical composition of Al-Mg-Si is different between these two cases, a simple comparison cannot be made, but in the UTPC processing of this study, after the solution temperature was reached and the cavitation stopped, specimens were rapidly cooled in water, similar to quenching. However, the decisive difference between natural aging by heat treatment and peening natural aging by UTPC is the presence of compressive residual stress after quenching of peening natural aging on the surface. When a GP zone is formed by natural aging, dislocations occur due to compressive residual stress, and the hardness increase is larger than that of natural aging by heat treatment. In fact, as shown in Figure 6 the hardness of the material processed by UTPC for 20 min increased by about 50 HV from 1 day to 9 days later.

The compressive residual stress after LTPC processing (20 MPa, 100 W, 20 min) was −170.6 MPa, and the compressive residual stress after UTPC processing (35 MPa, 800 W, 20 min) was −133.67 MPa [25]. Figure 7 shows the change in hardness over time after LTPC processing (20 MPa, 100 W, 20 min), which gives the highest compressive residual stress.

Unlike the case of UTPC, hardness did not increase over time. This is because the surface did not reach the solution temperature during LTPC processing, and peening natural aging did not occur. Similar to the results for UTPC processing in Figure 6, the material hardness of each LTPC processing time tended to decrease once and increase again. This is the result of a balance between changes in precipitates and the formation of new dislocations due to compressive stress after peening artificial aging, which has the highest compressive residual stress during processing, although it has not reached the solution temperature.

A qualitative SEM-EDS line analysis of the region of interest (ROI (Regions of Interest): the function that creates a window by specifying the X-ray energy range of interest in the spectrum) on the surface after UTPC and LTPC processing was carried out. Here, many-line qualitative analysis at 100,000 magnification was performed on the surface of specimens 1 day and 6 days after UTPC processing. Figure 8 shows the frequency distribution of the ROI analysis intensity of each Al, Mg, and Si concentration. The horizontal axis is the analytical intensity of various elements, and the vertical axis is the frequency distribution of the analytical intensity, showing the concentration distribution of various elements. It can also be recognized that the Mg concentration and Si concentration distribution were broader after 6 days in comparison with that after 1 day. When a GP zone is formed due to fluctuations in the concentrations of solute elements of Mg and Si by spinodal decomposition, the concentration of solvent elements changes to the same extent that the concentration of solute elements fluctuates. However, the fact that the spread of the Mg concentration distributions after 6 days of UTPC and 1 day of LTPC in Figure 8b are large, and the spread of solvent element Al concentration distributions are remarkably large in Figure 8a. This indicates that phase separation occurs due to spinodal decomposition in natural aging, which is consistent with the result of the increase in hardness in Figure 6. Furthermore, according to the result of qualitative analysis of ROI by SEM-EDS line analysis of the surface of a specimen 1 day after LTPC, the surface temperature did not reach the solution temperature. Therefore, phase separation occurred as it does without processing, and the solvent element Al concentration distribution is as wide as that in the specimen 6 days after UTPC.

Figure 9a,b shows the shape image and corrosion potential image of the mirror-polished surface before processing, respectively, as measured by KFM. The shape image is flat due to mirror polishing, and the potential image has no bias and a uniform potential distribution.

Figure 9c,d shows the KFM measurement results 72 days after UTPC processing of a specimen for 20 min; at this point, the hardness was starting to increase again. Although unevenness after processing is observed in the shape image, the potential image does not have a particularly biased distribution. In addition, the surface was leveled by raising the surface temperature to a level that generated solution treatment, and no remarkable peening marks were observed. Figure 9e,f provides the KFM measurement results after 104 days, when the hardness gradually increased after UTPC (WJ: 35 MPa, UC: 800 W, 60 min) processing. Although large irregularities are observed in the shape image, the potential image does not have a particularly biased distribution. In contrast, Figure 9g,h shows the KFM measurement results 7 days after LTPC processing of a specimen, which applies the highest compressive residual stress. A dent, which is considered to be a peening mark, is observed in the shape image. The surface potential at the peening mark is lower than that in the other regions. Further, the reason the peening marks were observed is that the surface did not reach the solution temperature and surface homogenization did not occur due to the lower processing temperature.

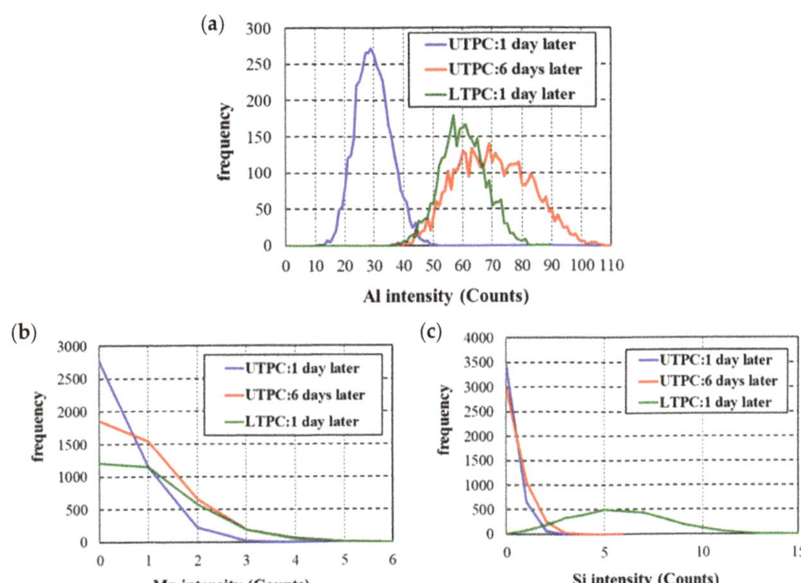

Figure 8. Frequency distribution of Mg and Si concentrations obtained from SEM-EDS line analysis 6 days after UTPC processing: (**a**) Al; (**b**) Mg; (**c**) Si. (WJ: 35 MPa, UC: 800 W, 20 min).

Table 2 gives the surface potential results for the various specimens. KFM can measure the work function, which is a measure of the ease with which electrons are emitted from the surface. Because oxidation involves the transfer of electrons, the higher the surface potential, the higher the corrosion resistance. Among aluminum alloys, AC4CH without processing has high corrosion resistance, and its surface potential after mirror polishing is as high as 1200 mV. The surface potential of special steel does not exceed 1000 mV, even after MFC processing [27]. Even if the aluminum alloy is mirror-polished, a passivation film is immediately formed and corrosion resistance is maintained. The surface potential of the specimen subjected to LPTC processing, which imparts the highest compressive residual stress, is almost the same as that of the as-received specimen after mirror polishing. UPTC processing, which causes peening natural aging, produced a decreased surface potential with 20 min of processing and increased potential with 30 min and 60 min of processing. After the outermost surface reaches the solution temperature and is subjected to the solution treatment by UTPC, precipitates such as GP zones are formed on the surface. The disparate surface potential results for the UTPC specimens are probably due to different formation states of these precipitates being created by different processing times (20 min vs. 30 min and 60 min). In the 60-min processed material, no change in surface potential was observed between 41 days and 104 days.

Table 2. Surface potential of aluminum alloy (AC4CH) by various processing.

Processing Condition	Mirror Polishing	LTPC	UTPC (1)	UTPC (2)	UTPC (3)	UTPC (4)
Elapsed days after processing	-	7	72	70	41	104
Water jet pressure (MPa)	-	20	35	35	35	35
Ultrasonic output (W)	-	100	800	800	800	800
Processing time (min)	-	20	20	30	60	60
Surface potential (mV)	1319	1256	879	2871	1762	1700

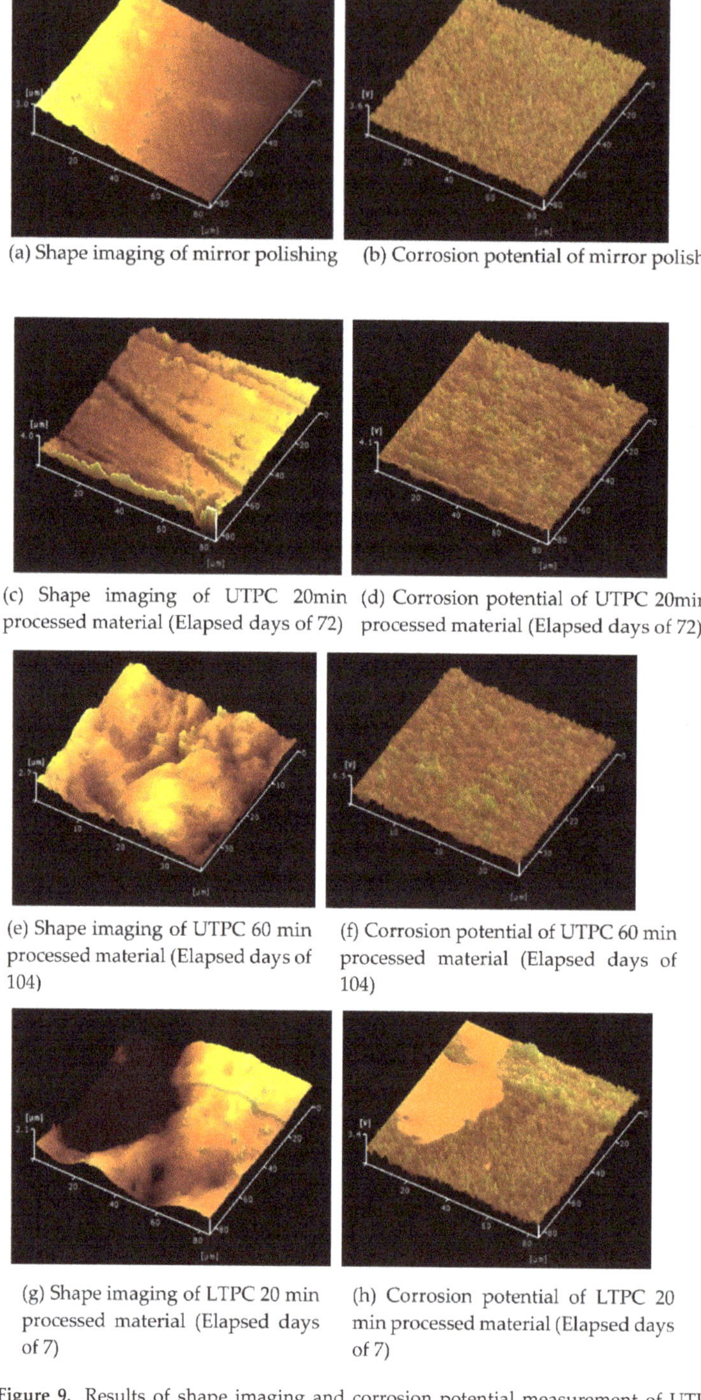

(a) Shape imaging of mirror polishing (b) Corrosion potential of mirror polishing

(c) Shape imaging of UTPC 20min processed material (Elapsed days of 72) (d) Corrosion potential of UTPC 20min processed material (Elapsed days of 72)

(e) Shape imaging of UTPC 60 min processed material (Elapsed days of 104) (f) Corrosion potential of UTPC 60 min processed material (Elapsed days of 104)

(g) Shape imaging of LTPC 20 min processed material (Elapsed days of 7) (h) Corrosion potential of LTPC 20 min processed material (Elapsed days of 7)

Figure 9. Results of shape imaging and corrosion potential measurement of UTPC and LTPC processed material by Kelvin force probe microscope (KFM).

In the UTPC 20-min processing and 30-min processing of the present study, the initial surface roughness was small (Ra: 0.1 μm); therefore, the surface was heated to the solution temperature, and the peening solution process was possible. In contrast, in UTPC 60-min processing, peening solution processing was performed up to 30 min; however, as the surface roughness increases with processing, the microjet impingement is not always vertical, and despite processing under UTPC conditions, UTPC processing does not occur. This is because the processing was performed at low temperature and low pressure. In addition, as can be seen from the shape image of Figure 9e,f due to long-term UTPC processing, high-temperature processing was not possible because the specific surface area increased and the ratio of heat dissipation to water increased rather than the heat input to the alloy interior. Therefore, it is probable that in the latter half of the 60-min processing, the peening artificial aging was performed by the pseudo-LTPC processing from the state where the solution was once processed. Therefore, as shown in Figure 6, hardening by natural aging did not occur after processing.

In UTPC and LTPC processing, the temperature of a metal surface is determined by the balance among heat input to the surface from a large number of cavitation bubbles with hot spots, convective heat dissipation from the surface to water, and heat dissipation due to heat conduction inside the metal. However, it is difficult to measure directly. Therefore, it is carried out by estimating the bubble temperature from the emission intensity (measurement of the number of photons per unit time) at the time of bubble collapse [9] and the temperature estimation is carried out by the structure change based on the equilibrium phase diagram after processing. For example, when UTPC processing is performed on Cr-Mo steel, it has been confirmed that spherical cementite that changes its structure at 500 °C or higher is formed just below the surface [28]. Therefore, it is considered that the surface temperature of AC4CH has reached the solution temperature in the UTPC treatment of this study.

Figure 10 shows the values of compressive residual stress applied to the surface immediately after UTPC and LTPC processing and after a long period of time. The compressive residual stress after 7 days of LPTC 20-min processed material is the highest at −171 MPa; however, after 141 days, it drops to the same level as UTPC 20-min processed material. However, the compressive residual stress of the UTPC 20-min processed material after 7 days is −134 MPa, which is not higher than that of the LPTC 20-min processed material, but the compressive residual stress does not decrease much even after 175 days. The compressive residual stress is reduced to −100 MPa or less in the UTPC 30-min processed material and UTPC 60-min processed material that have passed for a long time. The stability of the compressive residual stress of the UTPC peening natural aging material (UTPC 20 min) is superior to that of the LTPC peening artificial aging material (UTPC 20 min). With the artificial peening aging by LTPC, a compressive residual stress of −100 MPa or more can be applied even in a short time of 2 min, and as shown in Figure 7, a hardness of 120 to 140 HV can be obtained. Considering the decrease of compressive residual stress applied by the peening artificial aging for a long time, it can be judged that short-time processing is sufficient for peening artificial aging treatment by LTPC. Regarding the surface hardness, both the peening natural aging material by UTPC and the peening artificial aging material by LTPC reached almost 180 HV and achieved the same level of hardness after 100 days or more with longer-term processing of 20 min or more. From the viewpoint of corrosion resistance, the peening naturally aged material (UTPC 30 min) by UTPC is superior. From the above results, it is necessary to properly use peening natural aging and peening artificial aging depending on the intended use of the AC4CH alloy.

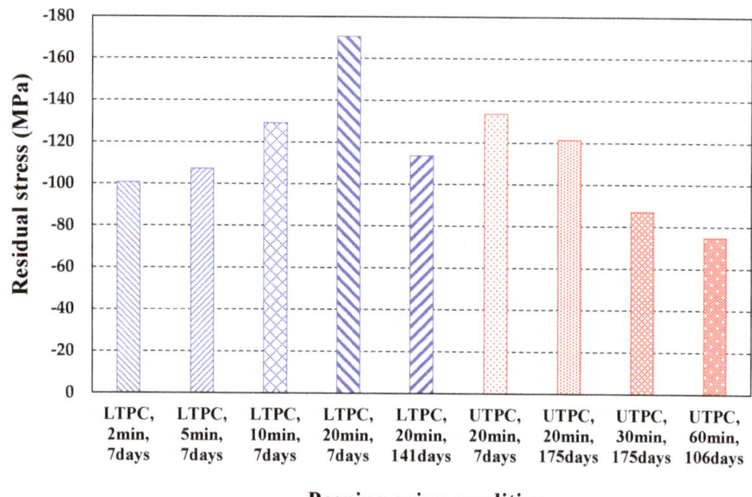

Figure 10. Aging variation of compressive residual stress of LTPC and UTPC.

4. Conclusions

Peening solution treatment was performed on AC4CH by UTPC processing to investigate peening natural aging. Furthermore, peening artificial aging treatment by LTPC was carried out to examine the change over time of peening natural aging and peening artificial aging, and the following findings were obtained:

(1) When the AC4CH alloy is processed for an appropriate time by UTPC processing, compressive residual stress is applied and natural aging occurs.

(2) The UTPC processing conditions of peening natural aging treatment to create a high surface compressive residual stress, surface hardness, and corrosion resistance were clarified.

(3) After artificial aging by LTPC treatment, the compressive residual stress decreases slightly with the passage of time, but the compressive residual stress is stabilized by peening natural aging by UTPC treatment.

(4) Processing for a short time does not cause peening natural aging by UTPC.

Author Contributions: All authors contributed to the paper either during the writing or editing phases. All authors made a substantial contribution to this research. Planning of research method and evaluation method, T.Y., M.I. (Masataka Ijiri) and S.K.; cavitation processing, M.I. (Masayoshi Iwamoto), T.Y. and M.I. (Masataka Ijiri); hardness measurement and surface roughness measurement, M.I. (Masayoshi Iwamoto); KFM measurement, T.O. and F.K.; measurement of compression residual stress, S.K. All authors have read and agreed to the published version of the manuscript.

Funding: This research was supported in part by JSPS KAKENHI Grant Number 19K04110 (Grant-in-Aid for Scientific Research (C)) and by the Light Metal Educational Foundation, Inc.

Data Availability Statement: Data is contained within the article. The data presented in this study are available on request from the corresponding author.

Conflicts of Interest: The authors declare no conflict of interest.

References

1. Summers, D.A. *Waterjetting Technology*; E&FN Spon: London, UK, 1995; ISBN 9780419196600.
2. Kim, Y.J.; Bae, D.H.; Kim, Y.J. The Improvement of High Cycle Fatigue Properties of AC4CH Alloy with Shot Peening Treatment. *Key Eng. Mater.* **2005**, *297-300*, 1919–1924.
3. Gao, Y.K.; Wu, X.R. Experimental investigation and fatigue life prediction for 7475-T7351 aluminum alloy with and without shot peening-induced residual stresses. *Acta Mater.* **2011**, *59*, 3737–3747. [CrossRef]

4. Kikuchi, S.; Nakamura, Y.; Nambu, K.; Ando, M. Effect of shot peening using ultra-fine particles on fatigue properties of 5056 aluminum alloy under rotating bending. *Mater. Sci. Eng.* **2016**, *A652*, 279–286. [CrossRef]
5. Hashish, M. Characteristics of surfaces machined with abrasive-waterjets. *J. Eng. Mater. Technol.* **1991**, *113*, 354–362. [CrossRef]
6. Hashish, M. A modeling study of metal cutting with abrasive waterjets. *J. Eng. Mater. Technol.* **1984**, *106*, 88–100. [CrossRef]
7. Yoshimura, T.; Tanaka, K.; Yoshinaga, N. Development of mechanical-electrochemical cavitation technology. *J. Jet Flow Eng.* **2016**, *32*, 10–16.
8. Yoshimura, T.; Tanaka, K.; Yoshinaga, N. Nano-level material processing by multifunction cavitation. *Nanosci. Nanotechnol. Asia* **2018**, *8*, 41–54. [CrossRef]
9. Yoshimura, T.; Maeda, D.; Ogi, T.; Kato, F.; Ijiri, M. Sonoluminescence from ultra-high-temperature and high-pressure cavitation and its effect on surface modification of Cr-Mo steel. *Global J. Technol. Optim.* **2020**, *11*. [CrossRef]
10. Summers, D.A.; Tyler, L.J.; Blaine, J.; Fossey, R.D.; Short, J.; Craig, L. Consideration in the design of a waterjet device for reclamation of missile casings. In Proceedings of the Fourth U.S. Water Jet Conference, University of California, Berkeley, CA, USA, 26 August 1987; pp. 82–89.
11. Gompf, B.; Gunther, R.; Nick, G.; Pecha, R.; Eisenmenger, W. Resolving sonoluminescence pulse width with time-correlated single photon counting. *Phys. Rev. Lett.* **1997**, *79*, 1405–1408. [CrossRef]
12. Yasui, K. Fundamentals of acoustic cavitation and sonochemistry. In *Theoretical and Experimental Sonochemistry Involving Inorganic Systems*; Ashokkumar, P.M., Ed.; Springer: Berlin/Heidelberg, Germany, 2011; Chapter 1; pp. 1–29.
13. Lorimer, J.P.; Mason, T.J. Applications of ultrasound in electroplating. *Electrochem* **1999**, *67*, 924–930. [CrossRef]
14. Peng, G.; Shimizu, S. Progress in numerical simulation of cavitating water jets. *J. Hydrodyn.* **2013**, *25*, 502–509. [CrossRef]
15. Yoshimura, T.; Tanaka, K.; Ijiri, M. Nanolevel surface processing of fine particles by waterjet cavitation and multifunction cavitation to improve the photocatalytic properties of titanium oxide. *IntechOpen Cavitation* **2018**. [CrossRef]
16. Vargel, C. *Corrosion of Aluminum, Chapter A.3—The Metallurgy of Aluminium*; Elsevier: Amsterdam, The Netherlands, 2004; pp. 23–57. [CrossRef]
17. Woo, W.; Choo, H.; Brown, D.W.; Feng, Z. Influence of the Tool Pin and Shoulder on Microstructure and Natural Aging Kinetics in a Friction-Stir-Processed 6061–T6 Aluminum Alloy. *Metall. Mater. Transac. A* **2007**, *38A*, 69–76. [CrossRef]
18. Fritsch, S.; Wagner, M.F.X. On the Effect of Natural Aging Prior to Low Temperature ECAP of a High-Strength Aluminum Alloy. *Metals* **2018**, *8*, 63. [CrossRef]
19. Rayleigh, L. On the pressure developed in a liquid during the collapse of a spherical cavity. *Philos. Mag.* **1917**, *34*, 94–98. [CrossRef]
20. Plesset, M.W. The dynamics of cavitation bubbles. *J. Appl. Mech.* **1949**, *16*, 277–282. [CrossRef]
21. Atchley, A.A. The Blake threshold of cavitation nucleus having a radius-dependent surface tension. *J. Acoust. Soc. Am.* **1988**, *85*, 152–157. [CrossRef]
22. Mitroglou, N.; Stamboliyski, V.; Karathanassis, I.K.; Nikas, K.S.; Gavaises, M. Cloud cavitation vortex shedding inside an injector nozzle. *Exp. Therm. Fluid Sci.* **2017**, *84*, 179–189. [CrossRef]
23. Lauterborn, W.; Kurz, T.; Geisler, R.; Schanz, D.; Lindau, O. Acoustic cavitation, bubble dynamics and sonoluminescence. *Ultrason. Sonochem.* **2007**, *14*, 484–491. [CrossRef] [PubMed]
24. Xiong, S.; Tandiono, T.; Ohl, C.D.; Liu, A.Q. Bubble pinch-off and breakup due to instability in microjetting. In Proceedings of the 17th International Conference on Miniaturized Systems for Chemistry and Life Sciences, Freiburg, Germany, 27–31 October 2013; pp. 71–73.
25. Yoshimura, T.; Ijiri, M.; Shimonishi, D.; Tanaka, K. Micro-forging and peening aging produced by ultra-high-temperature and pressure cavitation. *Int. J. Adv. Technol.* **2019**, *10*, 277. [CrossRef]
26. Hatta, H.; Tanaka, H.; Matsuda, S.; Yoshida, H. Effects of Mg, Si contents and natural aging conditions on the bake hardenability of Al–Mg–Si alloys. *J. Jpn. Inst. Light Metals* **2004**, *54*, 412–417. [CrossRef]
27. Ijiri, M.; Yoshimura, T. Sustainability of compressive residual stress on the processing time of water jet peening using ultrasonic power. *Heliyon* **2018**, *4*, e00747. [CrossRef] [PubMed]
28. Ijiri, M.; Yoshimura, T. Evolution of surface to interior microstructure of SCM435 steel after ultrahigh-temperature and ultra-high-pressure cavitation processing. *J. Mater. Process. Technol.* **2018**, *251*, 160–167. [CrossRef]

Review

Cavitating Jet: A Review

Hitoshi Soyama

Department of Finemechanics, Tohoku University, Sendai 980-8579, Japan; soyama@mm.mech.tohoku.ac.jp; Tel.: +81-22-795-6891; Fax: +81-22-795-3758

Received: 15 September 2020; Accepted: 15 October 2020; Published: 17 October 2020

Featured Application: Cavitation peening, Cleaning, Drilling.

Abstract: When a high-speed water jet is injected into water through a nozzle, cavitation is generated in the nozzle and/or shear layer around the jet. A jet with cavitation is called a "cavitating jet". When the cavitating jet is injected into a surface, cavitation is collapsed, producing impacts. Although cavitation impacts are harmful to hydraulic machinery, impacts produced by cavitating jets are utilized for cleaning, drilling and cavitation peening, which is a mechanical surface treatment to improve the fatigue strength of metallic materials in the same way as shot peening. When a cavitating jet is optimized, the peening intensity of the cavitating jet is larger than that of water jet peening, in which water column impacts are used. In order to optimize the cavitating jet, an understanding of the instabilities of the cavitating jet is required. In the present review, the unsteady behavior of vortex cavitation is visualized, and key parameters such as injection pressure, cavitation number and sound velocity in cavitating flow field are discussed, then the estimation methods of the aggressive intensity of the jet are summarized.

Keywords: cavitation; jet; vortex; mechanical surface treatment; cavitation peening

---

1. Introduction

Cavitation is a harmful phenomenon for hydraulic machineries such as pumps, as severe impacts are produced at bubble collapse [1,2]. However, cavitation impacts are utilized for mechanical surface treatment in the same way as shot peening, and this is named "cavitation peening" [3,4]. The great advantage of cavitation peening is that shots are not used in the peening process, as cavitation impacts are used in cavitation peening [5]. Thus, the cavitation-peened surface is less rough compared with the shot-peened surface, and the fatigue strength of cavitation peening is better than that of shot-peening [6]. In conventional cavitation peening, cavitation is generated by injecting high-speed water jet into water [3,4], and a submerged water jet with cavitation is called a "cavitating jet". The cavitation peening is utilized for the impacts of cavitation collapses, and it is different from water jet peening, in which water column impacts are used. To use the cavitating jet for peening, it is worth understanding the mechanism of the cavitating jet.

In a cavitating jet, cavitation is produced inside and/or outside a nozzle when sufficient pressure difference is applied across the nozzle. As Monkbadi reviewed, vortex structures in turbulent jets [7], ring vortices (0-mode), a single helical vortex (1st mode) and double-helical vortices (2nd mode) were observed in a cavitating jet [8–10]. In the case of a developed cavitating jet, which has been used for practical applications such as cutting, material testing, drilling and peening [11–17], cavitation clouds are shed periodically [16,18–27]. As it was reported that the lifetime of the cloud is a key factor in the aggressive intensity of the cavitating jet [28], the investigation of cavitation cloud shedding is very important. On the other hand, from the viewpoint of cavitation inception, a turbulent jet with cavitation was investigated [29–32]. In the present review, in order for the cavitating jet to be used for practical applications, the main subject is the developed cavitating jet, as shown in Figure 1a. In Figure 1, white bubbles are cavitation bubbles, as a used flash lamp was placed at the same side of a camera. As shown in Figure 1a, the cavitation clouds are clearly observed in the cavitating jet in water.

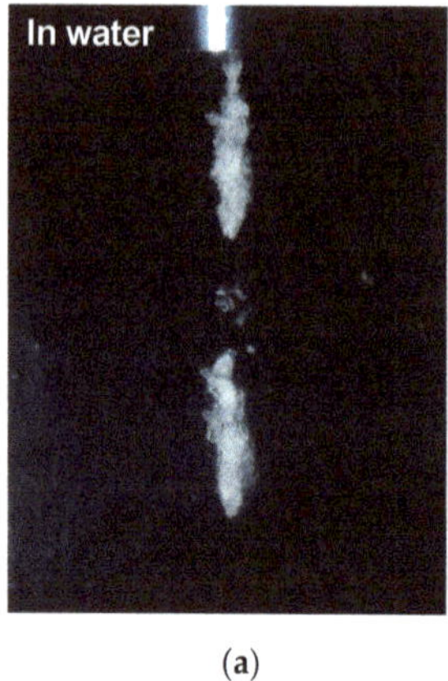

(a)

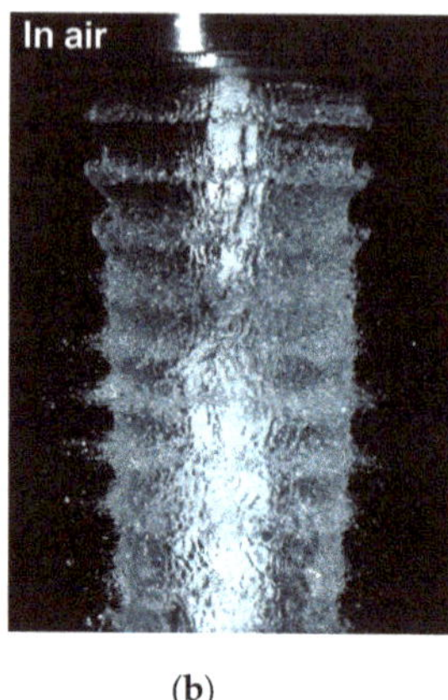

(b)

Figure 1. Typical aspects of cavitating jet: (**a**) cavitating jet in water; (**b**) cavitating jet in air.

Normally, a cavitating jet is produced by injecting a high-speed water jet into a water-filled chamber, as mentioned above. To apply a cavitating jet for components and/or plants which cannot be put in the chamber, Soyama developed a cavitating jet in air by injecting a high-speed water jet into a low-speed water jet, which was injected into air without the water-filled chamber using a concentric nozzle [33–36]. A typical cavitating jet in air by injecting a high-speed water jet into a low-speed water jet is shown in Figure 1b. Cavitation clouds are also observed in the water column of low-speed water jets of cavitating jets in air, as shown in Figure 1b. Even though the cavitating jet was in air, cloud shedding was observed [35,37,38]. In addition, in the case of the cavitating jet in air at optimum conditions, the wavy pattern of the low-speed water jet was observed, as shown in Figure 1b [34,35,37,38]. The frequency of the wavy pattern is equal to the shedding frequency of the cloud [35]. Even for the cavitating jet in air, unsteady behavior is very important.

In cavitating jets both in water and in air, whereas the cloud shedding frequency is several hundred Hz [18,35,39,40], only a few severe impacts occur per second [34,41–43] when cavitation impacts are measured by special-made PVDF transducers [41,42]. Thus, in order to enhance the aggressive intensity of the cavitating jet in water and air for practical applications of cavitating jets, the unsteady behavior of the cavitating jet should be investigated. In the present review, the normal cavitating jet, i.e., the cavitating jet in water, was mainly discussed, and the cavitating jet in water was simply described as the cavitating jet.

To simulate the cavitating jet numerically, it is important to understand the flow pattern of the cavitating jet. As is well known, numerical simulation of cavitating flow is not yet easy, due to the high Reynolds number and phase change phenomenon. Numerical investigation of three-dimensional cloud cavitation with special emphasis on collapse induced shock dynamics [44], simulation of cloud cavitation on propeller [45], and the hydrofoil [46] were carried out. In the area of numerical simulation of the cavitating jet, considering bubble dynamics, the numerical simulation of vortex cavitation in a three-dimensional submerged transitional jet near inception condition was carried out [47], and bubble growth in the shear layer of the cavitating jet was calculated [48]; residual stresses introduced by cavitating jet considering bubble growth and collapse were also simulated [49]. The cavitating flow in a venturi nozzle was also tried by a large eddy simulation of turbulence–cavitation interactions [50]. From the viewpoint of cloud-shedding of the cavitating jet, numerical analysis of cavitation cloud-shedding in a free submerged water jet was carried out [51,52].

The interaction of cavitation bubbles and materials was also investigated numerically [53–56]. In the case of the aggressive cavitating jet, the clouds shed periodically and form ring vortex cavitation on the impinging surface, and then collapse, producing severe impacts. Thus, the experimental investigation of the flow pattern would be assisted by numerical simulations for more precise investigations of the cavitation impacts produced by the cavitating jet.

In the present review, in order to use the cavitating jet for practical applications, papers that investigated the flow pattern of the cavitating jet experimentally were reviewed. To enhance the aggressive intensity of the cavitating jet for these applications, the key factors of the cavitating jet were also summarized, and the estimation method of the aggressive intensity of the jet was discussed. In the present paper, the aggressive intensity of the cavitating jet means erosion rate measured by weight loss of the target metals and/or the peening intensity measured by the arc height of the metallic plate.

2. Cavitation

Cavitation is a phase-change phenomenon in which the liquid phase is changed to a gas phase due to a decrease in liquid pressure to vapor pressure by increasing flow velocity [2], and it is called "hydrodynamic cavitation". Cavitation is also generated by ultrasonic vibration, which is named "ultrasonic cavitation". Bubbles are also formed by irradiating a pulse laser into water, in which the bubble behaves as a cavitation bubble [57]; this is called "laser cavitation". A schematic diagram of cavitation is illustrated in Figure 2a. When a cavitation nucleus such as a tiny air bubble is subjected to a low-pressure region, it becomes a cavitation, and it develops and shrinks, then collapses, producing microjet and shock wave during rebound. The microjet and the shock wave produce a severe impact, which can deform metals. After rebound and shrinking, bubbles remain in the water, which is called a residual bubble [58]. In the research area of cavitation, including bubble dynamics, spherical bubbles have mainly been investigated by numerical simulations [59] and experimental studies [57,60,61], and the effects of bubble shape and bubble interactions have also been studied [62].

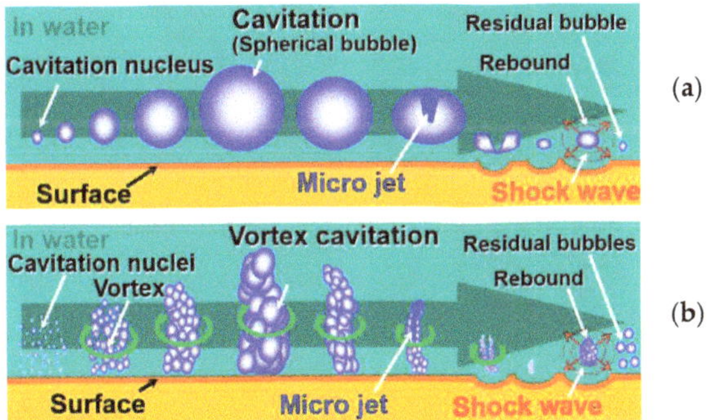

Figure 2. Schematic diagram of development and collapse of cavitation: (**a**) spherical bubble; (**b**) vortex cavitation.

In experimental studies of severe cavitation erosion, it was found that vortex cavitation, as shown in Figure 3 [63], produced severe impacts on fluid machineries such as pumps and valves [1,64,65]. A schematic diagram of vortex cavitation is shown in Figure 2b. Cavitation nuclei accumulate in a vortex, and they become vortex cavitation in a high-speed region, i.e., a low-pressure region. The vortex cavitation develops and shrinks, then collapses. Regarding a model test of vortex cavitation using a rotating device [66], a microjet was observed in the vortex cavitation. Thus, in order to use cavitation impacts for practical applications, the generation of vortex cavitation is very important.

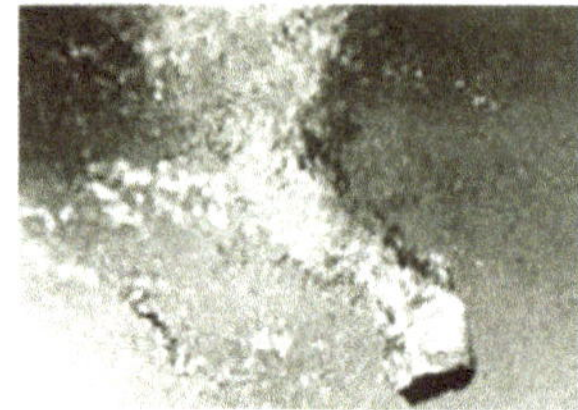

Figure 3. Typical aspect of vortex cavitation, which produces severe impact [63].

3. Cavitating Jet

3.1. Structure of Cavitating Jet

To show the flow pattern of the cavitating jet, Figure 4 illustrates aspects of the impinging cavitating jet taken with (a) a flash lamp of 1.1 μs and (b) a shutter speed of 1/25 sec, i.e., 40 ms. Thus, Figure 4a reveals an instantaneous aspect of the jet, and Figure 4b shows a kind of time-averaged aspect of the impinging cavitating jet. In the nozzle used for the cavitation jet, the cavitator and the guide pipe were installed to enhance the aggressive intensity of the cavitating jet [39], and standoff distance was defined as the distance from the upstream corner of the nozzle to the specimen surface. As shown in Figure 4a, a cloud cavitation was observed between the nozzle and the impinging surface, and a ring vortex cavitation was observed on the surface. When the cavitating jet was observed by a normal light, as shown in Figure 4b, the cavitating jet seems to be a continuous jet from a nozzle to an impinging surface. However, cloud cavitation sheds from the nozzle to the surface, and the cloud cavitation becomes a ring vortex cavitation.

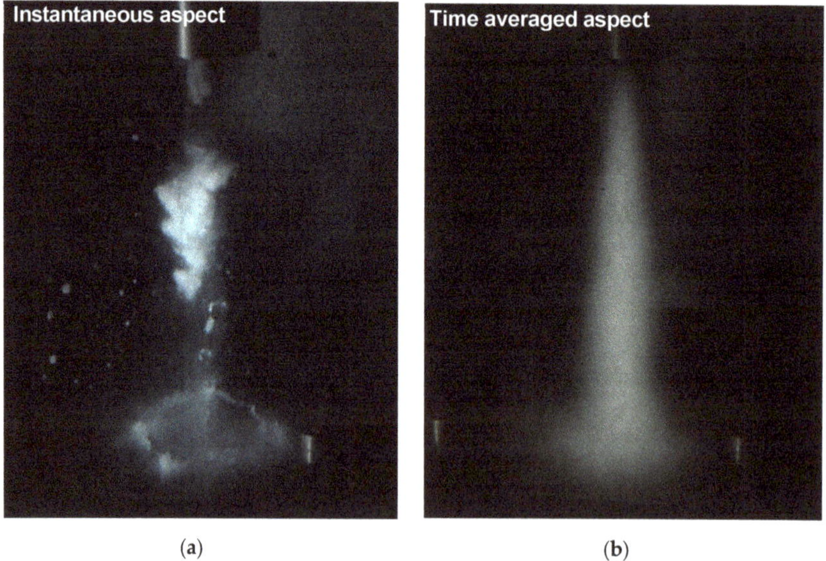

Figure 4. Aspects of cavitating jet (cylindrical nozzle, nozzle diameter d = 2 mm, upstream pressure p_1 = 30 MPa, downstream pressure p_2 = 0.1 MPa, standoff distance s = 222 mm): (**a**) observation with flash lamp whose exposure time is 1.1 μs; (**b**) observation with normal light and shutter speed of 40 ms.

To show vortex cavitation in the cavitating jet more precisely, Figure 5a reveals the aspect of the free cavitating jet through a conical nozzle, and Figure 5b shows the aspect of the impinging cavitating jet with a flash lamp [9]. As shown in Figure 5b, the PVDF transducers [41,42] were installed in the impinging surface, which the white arrow shows, and the signal from the PVDF transducer was synchronized to the flash lamp; thus, Figure 5b reveals the aspect of the jet which produces the impact on the impinging surface. As shown in Figure 5a, helical vortex cavitation is observed near the nozzle outlet, and cloud cavitation is observed downstream from the nozzle. Thus, the cloud cavitation results from merging the vortex cavitation. In the case of the impinging cavitating jet, a part of the ring cavitation collapses on the surface producing the impact, which was detected by the PVDF transducer. The ring vortex cavitation produces a severe impact on the impinging surface. In view of the practical applications of cavitation impacts generated by the cavitating jet, the collapse of the ring vortex cavitation is an important phenomenon.

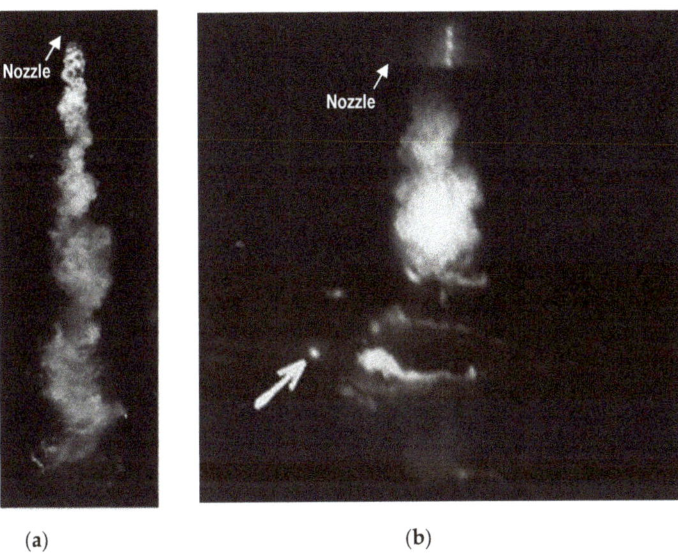

(a) (b)

Figure 5. Vortex cavitation in cavitating jet: (**a**) helical vortex cavitation in free cavitating jet (conical nozzle, nozzle diameter d = 2.1 mm, upstream pressure p_1 = 6 MPa, downstream pressure p_2 = 0.18 MPa); (**b**) ring vortex cavitation on impinging surface (cylindrical nozzle, nozzle diameter d = 2 mm, upstream pressure p_1 = 12 MPa, downstream pressure p_2 = 0.36 MPa, standoff distance s = 40 mm) [9].

To reveal periodical shedding of cloud cavitation from the nozzle, as mentioned in the introduction, Figure 6 shows the aspect of the impinging cavitating jet taken by a high-speed video camera. The nozzle shown in Figure 6 was the same nozzle as in Figure 4, and it had the cavitator and the guide pipe. At t = 0 ms, the cavitation cloud sheds from the nozzle to the downstream, and the cloud reaches the impinging surface at t = 1.5 ms. A part of cloud cavitation becomes a ring vortex cavitation at t = 1.75 ms, and it spreads out on the surface and then collapses. Although the water jet is injected continuously into water, the vortex cavitations near nozzle shed downstream coalescing each other and they become a large cloud cavitation, thus the bubble density between clouds is reduced. At t = 4.25 ms, the new cavitation cloud, whose shape is similar to that at t = 0 ms, sheds from the nozzle. Thus, it is a periodical phenomenon, as previously reported [16,18–27].

According to a previous report [28], as shown in Figure 7, some cloud, i.e., the 1st cloud in Figure 7a, stays near the nozzle, and the 2nd jet core passes through the 1st cloud, as shown in Figure 7b,c; then, the 2nd jet core produces the 2nd cloud downstream of the 1st cloud. The 3rd jet core passes through the 1st and 2nd clouds and produces the 3rd cloud downstream of the 2nd cloud, as shown in Figure 7d. The 5th jet core produces the 5th cloud downstream of the 1st cloud as shown in Figure 7e. Detailed images of the free cavitating jet observed by the high-speed video camera and the schematic cross-sectional diagrams of the progress of a cavitating jet are shown in reference [28]. In Figure 6, the cavitation cloud marked by the yellow arrow stays at nearly the same position. The cloud cavitation staying where it is means that it has a long lifetime. The jet core with a longer lifetime cavitation cloud impinges the surface with high impingement pressure, as a drag of the jet core with the larger cloud is lesser than that of a smaller cloud because of the density of water and bubbles. Thus, the aggressive intensity of the cavitating jet with a longer lifetime cavitation cloud is larger.

Namely, the lifetime of the cloud is a key factor of the cavitating jet, and the longer the cloud lifetime, the more aggressive the intensity of the cavitating jet [28].

Figure 8 shows a typical aspect of a pure aluminum specimen that was exposed to the fixed cavitating jet, revealing the treatment area of the fixed cavitating jet. Plastic deformation pits are observed in a ring region, whose outer diameter is 60 mm and inner diameter is 30 mm. When the nozzle is scanned or the specimen is moved, the treatment area is uniform [67]. In Figure 8, as the nozzle throat diameter was 2 mm, the cavitating jet can treat an area 30 times wider than that of the nozzle throat. While the white region, which was impinged by the jet core, was observed at the jet center, the main treatment area is the ring region. Namely, the jet center is not treated by cavitation impacts. The understanding of the treatment area by the cavitating jet shows that a ring region is very important when using the cavitating jet for the practical applications.

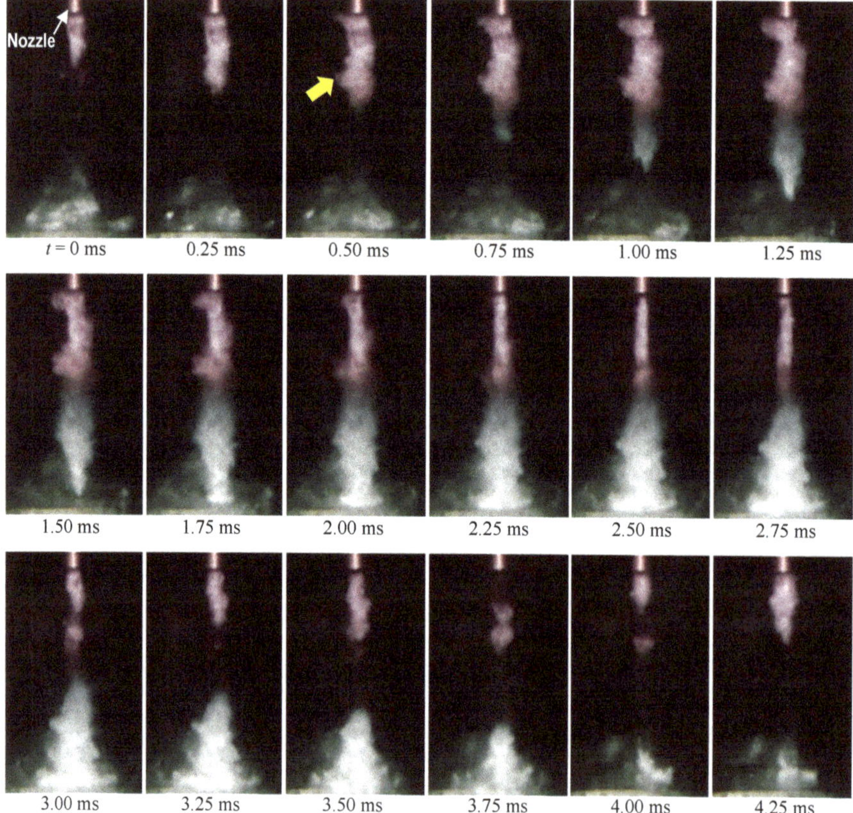

Figure 6. Aspects of impinging cavitating jet observed by high-speed video camera (cylindrical nozzle, nozzle diameter d = 2 mm, upstream pressure p_1 = 30 MPa, downstream pressure p_2 = 0.1 MPa, standoff distance s = 222 mm).

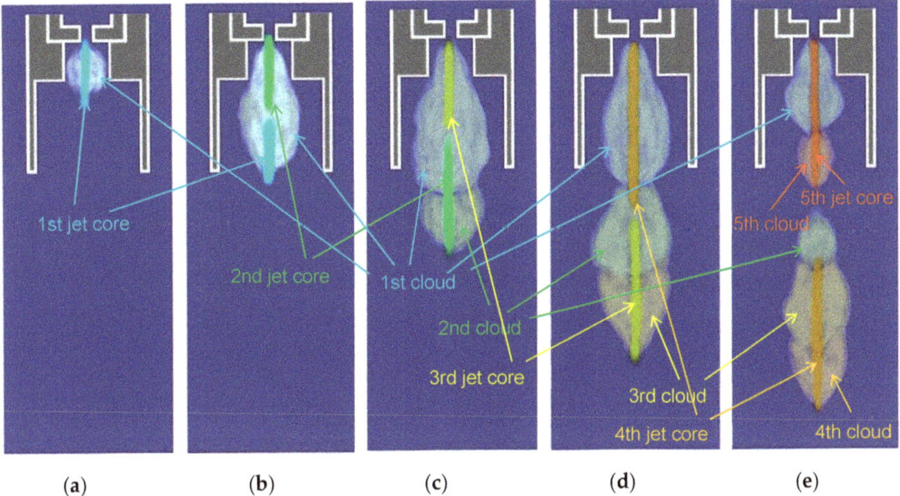

Figure 7. Schematic diagram of the development of cavitation cloud in cavitating jet: (**a**) 1st cloud produced by the 1st jet core; (**b**) 2nd jet core passing through the 1st cloud; (**c**) 2nd cloud produced by the 2nd jet core and 3rd jet core passing through the 1st cloud; (**d**) 3rd cloud produced by the 3rd jet core and 4th jet core passing through the 1st and 2nd clouds; (**e**) 4th cloud produced by the 4th jet core, and 5th cloud produced by 5th jet core passing through the 1st clouds.

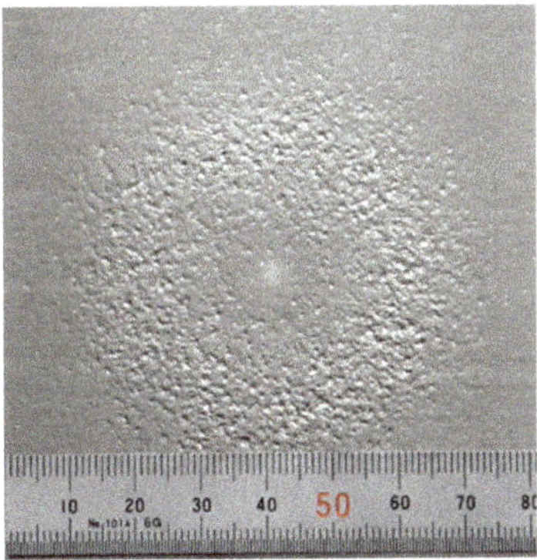

Figure 8. Typical treatment area by a fixed cavitating jet (pure aluminum, nozzle diameter d = 2 mm, upstream pressure p_1 = 30 MPa, downstream pressure p_2 = 0.1 MPa, standoff distance s = 262 mm, exposure time t = 1 min) [4].

Regarding the reason for the ring region, Figure 9 illustrates the schematic diagram of the impinging cavitating jet, considering the observations of the cavitating jet by the instantaneous photograph and the high-speed video. As shown in Figure 4b, when the cavitating jet is observed by an instantaneous photograph with normal light, the cavitating region seems to be a continuous jet, as shown in Figure 9a. It was previously thought that swirl cavitation in the shear region around the jet directly hits the impinging surface and that the swirl directly produces a ring treatment area [68–71]. However, this is incorrect, because the aspect of the cavitating jet, as shown in Figure 4b is a kind of time-averaged cavitating region. Considering Figures 4a, 5 and 6, vortex cavitation is initiated in and near the nozzle outlet, and cloud cavitations combine each other; then, the cloud cavitation forms the ring vortex cavitation on the impinging surface. Thus, in order to simulate bubbles in the cavitating jet, the pressure hysteresis of these processes should be considered. For example, if the cavitation in swirl directly hit the impinging surface, the pressure around cavitation would be the pressure in the shear layer around the jet, and it gradually increases with the shedding of cavitation, as the jet speed decreases from the distance from the nozzle. On the other hand, the pressure of the cloud cavitation, which impinges the surface, increases at the impinging jet center suddenly and decreases in the ring vortex cavitation, then increases again.

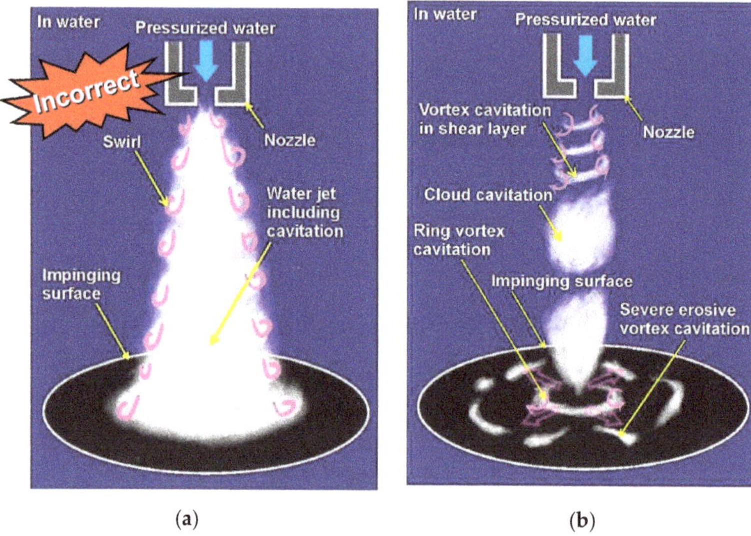

Figure 9. Schematic diagram of cavitating jet: (**a**) swirl; (**b**) vortex, cloud and ring cavitation.

To consider the mechanism of the ring erosion occurring on the impinging surface, Figure 10 shows a schematic diagram of the local cavitation number on the surface [72]. The impinging pressure profile on the surface has a maximum at the jet center, p_{max}, and it changes with the injection pressure and cavitation number. The b in Figure 10 is a kind of jet width defined by the flow velocity [72]. The cavitation number of the cavitating jet, σ, is defined by the injection pressure, i.e., the upstream pressure of the nozzle p_1, the downstream pressure of the nozzle p_2 and the vapor pressure of water p_v as follows [2], and is simplified as Equation (1) and as $p_1 >> p_2 >> p_v$.

$$\sigma = \frac{p_2 - p_v}{p_1 - p_2} \approx \frac{p_2}{p_1} \tag{1}$$

In Figure 10, the local cavitation number σ_L is defined by Equation (2) and is proportional to the ratio of $p - p_v$ and $\frac{1}{2}\,\rho\, v_{max}{}^2$ [72].

$$\sigma_L \propto \frac{p - p_v}{\frac{1}{2}\,\rho\, v_{max}{}^2} \propto \frac{p - p_v}{p_{max} - p_2} f(r) \tag{2}$$

Here, p and v_{max} are the pressure and the maximum flow velocity on the impinging surface. The v_{max} has a maximum at a certain distance from the jet center, r, as it is zero at the jet center and at a further point from the jet center as shown in the lower figure of Figure 10. As v_{max} is mainly determined by the pressure difference, $p_{max} - p_2$, σ_L is described by a function of $f(r)$, as shown in the right-hand term of Equation (2). When the ring vortex cavitation sheds on the surface, the cavitation develops the $\partial\sigma_L/\partial r < 0$ region, and it collapses at $\partial\sigma_L/\partial r \approx 0$. This is why the ring treatment area is obtained by the impinging cavitating jet. The detail of the pressure distribution on the impinging surface was shown in references [72,73].

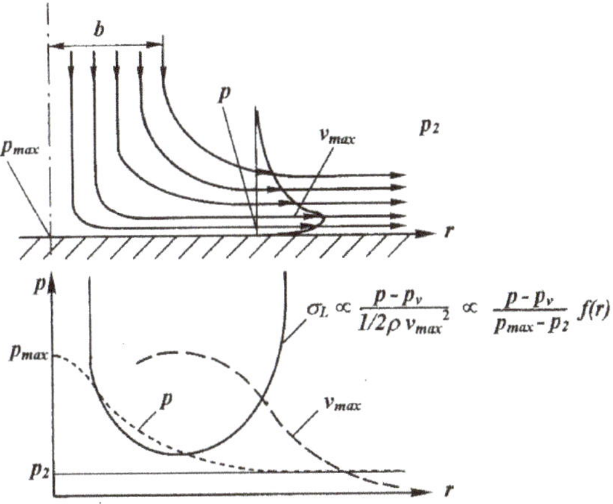

Figure 10. Schematic diagram of local cavitation number [72].

3.2. Periodical Shedding of Cavitation Cloud

As mentioned above, the cavitation cloud sheds periodically [16,18–27], and it forms the ring vortex cavitation; then, the ring vortex cavitation collapses, producing impacts. Thus, the periodical shedding of the cavitation cloud is an important phenomenon. Figure 11 reveals aspects of periodical shedding of the cavitation cloud of a cavitating jet [21]. In Figure 11, the jet flows from the left-hand side to the right-hand side. As shown in Figure 11a, the cavitation cloud breaks near the nozzle at 0.5, 0.9, 1.4 and 1.9 ms; thus, the shedding frequency f_{shedd} is about 2 kHz for p_1 = 20 MPa and σ = 0.014. When p_1 is increased to 30 MPa at a constant cavitation number, i.e., σ = 0.014, f_{shedd} is increased to 2.4 KHz, as shown in Figure 11b. When σ is increased from 0.014 to 0.02 at constant p_1= 20 MPa, f_{shedd} also increases from 2 to 3.2 KHz, as shown in Figure 11a,c. f_{shedd} is affected by not only p_1 but also σ. As is well known, the cavitating length L_{cav} and the width w_{cav} are affected by p_1, σ and d. In a previous report, f_{shedd} was measured experimentally, changing with p_1, σ and d, and the following relations were found [21].

$$f_{shedd} \propto w^{-1} \tag{3}$$

$$f_{shedd} \propto p_1{}^{0.45\pm0.03} \tag{4}$$

$$f_{shedd} \propto d^{-0.98\pm0.14} \tag{5}$$

$$f_{shedd} \propto \sigma^{0.83\pm0.10} \tag{6}$$

When the Strouhal number, St, is defined by f_{sedd}, w and jet velocity at the nozzle exit, U, which is calculated from p_1, the following experimental equation is obtained [21].

$$St = \frac{f_{shedd} \cdot w}{U} \approx 0.18 \pm 0.02 \tag{7}$$

This result suggests that the cavitation cloud shedding is a phenomenon governed by a constant Strouhal number. On the other hand, the nozzle outlet geometry affects the aggressive intensity of the cavitating jet [39,74]. Details are described in Section 4.6. The f_{sedd} is also affected by the nozzle outlet geometry. For example, when the guide pipe with the cavitator was installed, the aggressive intensity of the jet was four times larger than that without the guide pipe and the cavitator [39], and St became nearly a quarter of that of the jet without the guide pipe and the cavitator [75], as f_{sedd} was decreased. Namely, the constant value of St should be unique to the nozzle outlet geometry. The investigation of St for various nozzles would be a future work, to enhance and/or control the aggressive intensity of the cavitating jet.

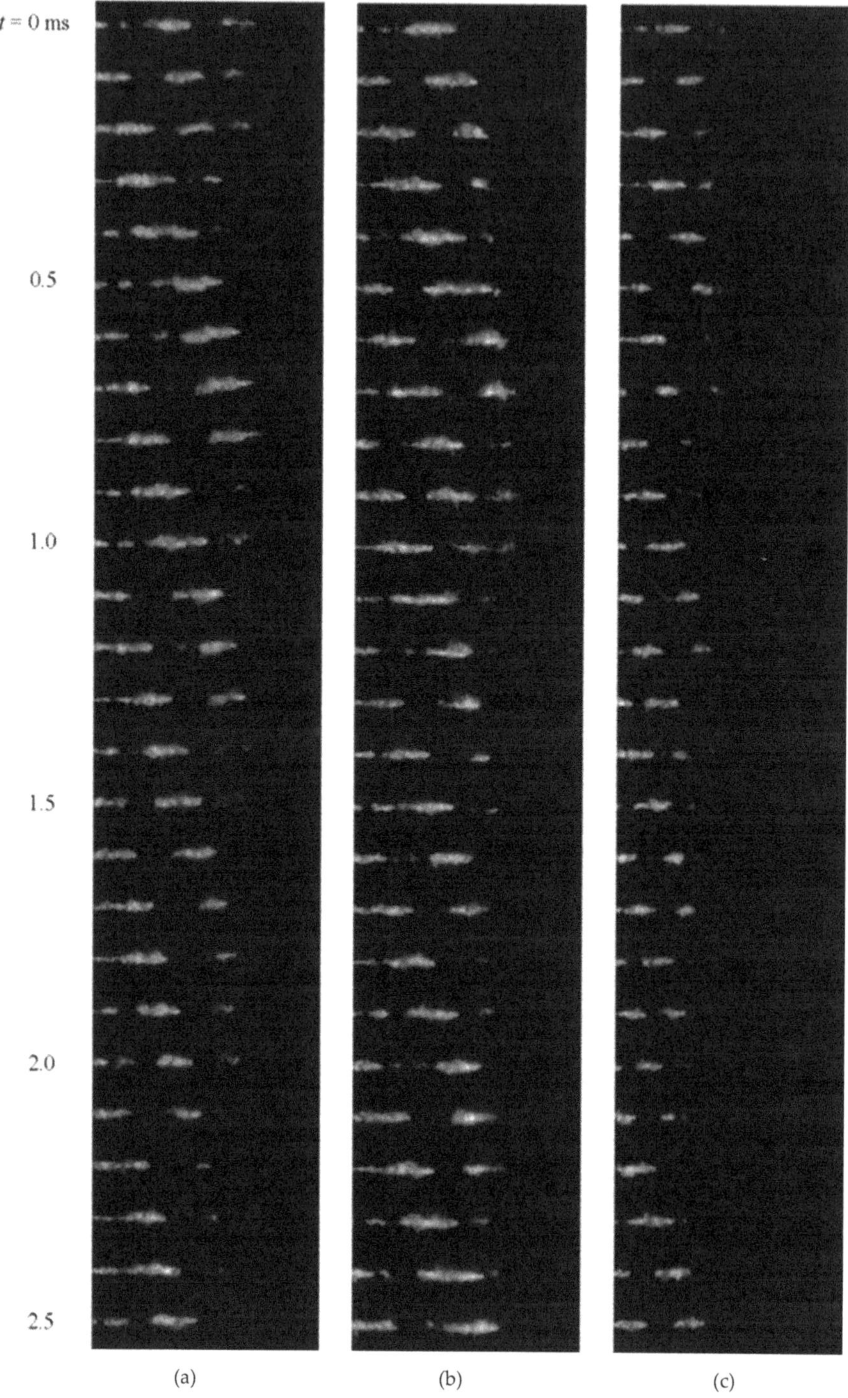

Figure 11. Aspects of periodical shedding of cloud cavitation: (**a**) $p_1 = 20$ MPa, $\sigma = 0.014$; (**b**) $p_1 = 30$ MPa, $\sigma = 0.014$; (**c**) $p_1 = 20$ MPa, $\sigma = 0.02$ [21].

4. Key Parameters of Cavitating Jet

4.1. Type of Cavitating Jet

As mentioned above, a cavitating jet normally means a submerged high-speed water jet with cavitation, i.e., a cavitating jet in water. Soyama developed the cavitating jet in air by injecting a high-speed water jet into a low-speed water jet without a water-filled chamber [33,35], for cavitation peening treatment outside of a tank and/or pipes [34,76]. When the residual stress of stainless steel was measured, it was reported that the cavitating jet in water introduced compressive residual stress in a deeper region, and the cavitating jet in air introduced large but shallow compressive residual stress on the surface [4]. The cavitation peening using the cavitating jet in water corresponds to shot peening using large shots, and that of the cavitating jet in air corresponds to shot peening using small shots at high velocity. Thus, the characteristics of the treated surface strongly depend on the type of the cavitating jet.

4.2. Standoff Distance

As shown in Figure 9b, as cavitation is generated inside and/or outside of the nozzle, it becomes cloud cavitation and forms ring vortex cavitation on the impinging surface, then collapses, producing the impacts. In view of the practical applications of the cavitating jet, the working mechanism strongly depends on the standoff distance, which is the distance from the nozzle to the surface. Figure 12 illustrates the weight loss as a function of standoff distance at constant injection pressure, p_1 = 120 bar (12 MPa) [77]. In Figure 12, the weight loss means a kind of aggressive intensity of the jet. Two peaks are observed at each condition in Figure 12. For convenience, the peak near the nozzle side and the peak at the further nozzle side are named the 1st peak and 2nd peak, respectively. The 1st peak results from the impacts produced by water column collisions in the jet center. Even for the submerged water jet, the similar mechanism of water jet cutting is still active, whereas the affective region is limited near the nozzle. The 2nd peak is generated by the cavitation impact, as shown in Figures 6 and 9b. As cavitation is developed and then collapsed, a certain distance from the nozzle is required. As shown in Figure 12, the optimum standoff distance s_{opt} of the 1st peak is scarcely affected by cavitation number, and that of the 2nd peak strongly depends on cavitation number. At p_2 = 2.4 bar (0.24 MPa), the weight losses at 1st peak and 2nd peak are 250 and 450 mg, respectively. They are 250 and 220 mg for p_2 = 3.0 bar (0.3 MPa) and 300 and 110 mg for p_2 = 3.6 bar (0.36 MPa). When the maximum values of the 1st peak and 2nd peak are compared, the value of the 2nd peak at p_2 = 0.24 MPa is 1.5 times larger than that of the 1st peak at p_2 = 0.36 MPa. Namely, at optimum cavitating conditions, the aggressive intensity due to cavitation impact, i.e., the 2nd peak, is larger than that of water column impact, i.e., 1st peak. Note that water jet peening and cavitation peening use the 1st peak and 2nd peak, respectively.

To avoid confusing cavitation peening and water jet peening, Soyama proposed a classification map for cavitation peening and water jet peening using standoff distance and cavitation number, as shown in Figure 13. Over 150 points were collected from references [39,74,77–89], and it was found that the line shown in Equation (8) distinguished between cavitation peening and water jet peening [90].

$$\frac{s_{opt}}{d} = 1.8\,\sigma^{-0.6} \tag{8}$$

One easy way to confirm the 2nd peak region, i.e., cavitating peening region, is as follows. Considering the aggressive intensity of the cavitating jet, the target materials are chosen. For example, in the case of the weak cavitating jet, a soft metal would be better. Then, the target is exposed to the jet. When a ring pattern is obtained, such as in Figure 8, it is a cavitation peening condition. The removal of paint can show the treatment area of cavitation peening [91]. A pressure-sensitive film also detects the treatment area of a cavitating jet [12].

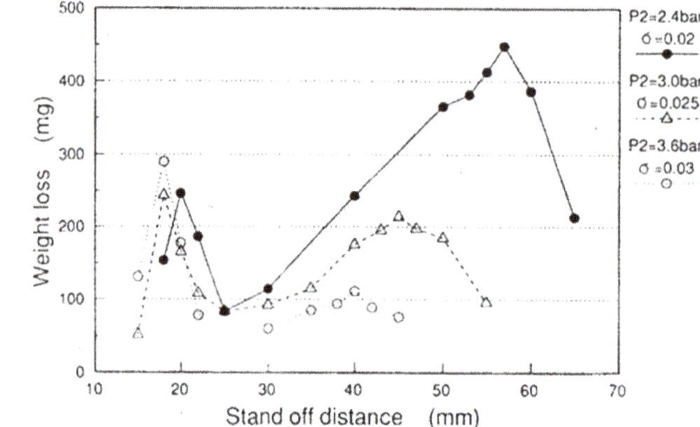

Figure 12. Weight loss, i.e., the aggressive intensity of the jet, as a function of standoff distance changing with cavitation number at constant injection pressure. The peak at the near side of the nozzle, i.e., the 1st peak, was caused by water column impacts. The peak at the far side from the nozzle, i.e., the 2nd peak, was produced by cavitation impacts. The standoff distance of 2nd peak was changed by the cavitation number [77].

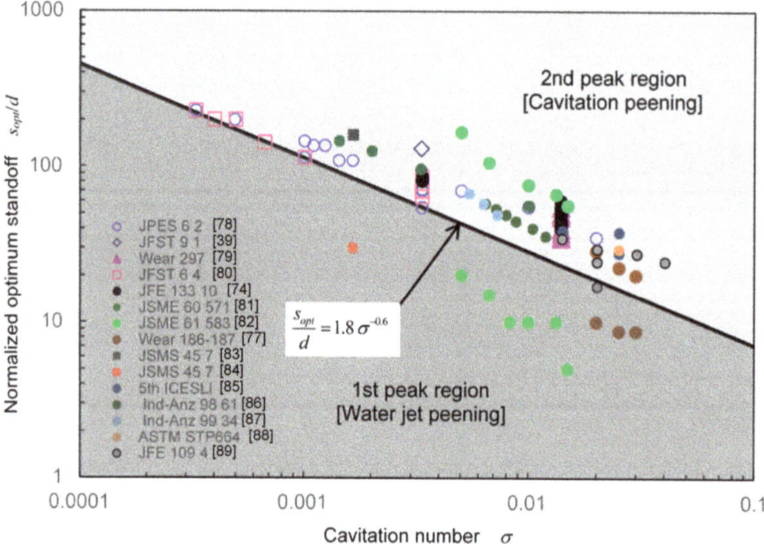

Figure 13. Classification map for cavitation peening and water jet peening considering standoff distance and cavitation number [90].

4.3. Injection Pressure

To show the effect of injection pressure on the processing capability of the cavitating jet, Figure 14 reveals the processing capability β at (a) the 1st peak, i.e., water jet peening, and (b) 2nd peak, i.e., cavitation peening, at the constant downstream pressure condition [92]. The processing capability β is defined by the arc height h of band steel made of the same material as Almen strip, considering the width of the steel w_s and the peening width w_p, as follows [92].

$$\alpha = \left| 1 - \frac{w_p}{w_s} \right| \tag{9}$$

$$\frac{1}{\rho} = \frac{8h}{L^2} \quad (10)$$

$$\beta = \frac{1}{\rho}(1+\alpha) \quad (11)$$

Here, L is the length to measure h. In Figure 14, the processing capability, i.e., a kind of aggressive intensity, was obtained by the arc height of the peened plate, as the arc height using the Almen strip is commonly used to measure the peening intensity [93]. In the case of water jet peening, β increases with the injection pressure p_1, as the peening effect is produced by water column impacts, which increases with p_1. On the other hand, in the case of cavitation peening, β has a maximum at p_1 = 40 MPa at a constant downstream condition. When the maximum values of the 1st peak and 2nd peak are compared: β at the 2nd peak is 1.7 times larger than that at the 1st peak. As the jet power of 60 MPa is 1.8 times larger than that of 40 MPa, the peening efficiency of cavitation peening is about three times higher than that of water jet peening. Note that too high an injection pressure reduces the peening intensity of cavitation peening, as shown in Figure 14b. The reason the peening intensity of cavitation peening decreases at p_1 > 40 MPa is discussed in "5. Estimation of Aggressive Intensity of Cavitating Jet".

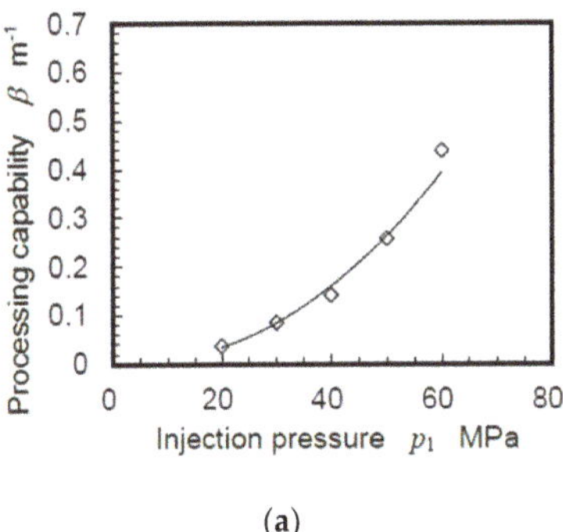

(a)

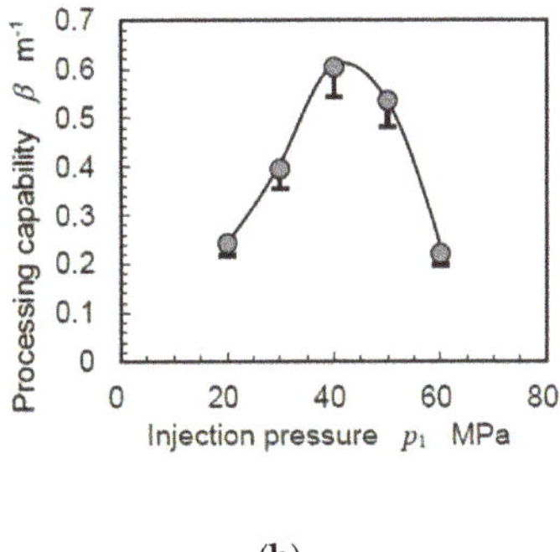

(b)

Figure 14. Effect of injection pressure on the aggressive intensity of cavitating jet at constant downstream pressure condition; (**a**) 1st peak; (**b**) 2nd peak [92].

4.4. Cavitation Number

As the cavitating jet is the cavitating flow, the cavitation number is one of the key parameters of the cavitating jet. To reveal how important the cavitation number is in the cavitating jet characteristics, the relationship between the cavitation number and the optimum standoff distance is shown in Figure 15 [72]. In Figure 15a, the cavitating length is also added, and the length and the distance are normalized by effective nozzle diameter D_e, which is defined by the nozzle throat diameter and discharge coefficient [72]. The data of the references are put in together in Figure 15 [77,81,86,87,89,94], as well as the relationship on each line, which is a straight line on a log–log scale. That is, the relationship can be described by Equation (12) [72].

$$\frac{s_{opt}}{d} = c_1 \sigma^{-c_2} \quad (12)$$

Here, c_1 and c_2 are constants, and they depend on the nozzle geometry.

In Figure 15b, the expected standoff distance considering the pressure distribution and the local cavitation number on the impinging surface is also shown. The expected standoff distance is on the line of the log–log scale, and it is very close to the experimental result. The local cavitation number on the impinging surface is also an important parameter when considering the flow pattern of the impinging cavitating jet.

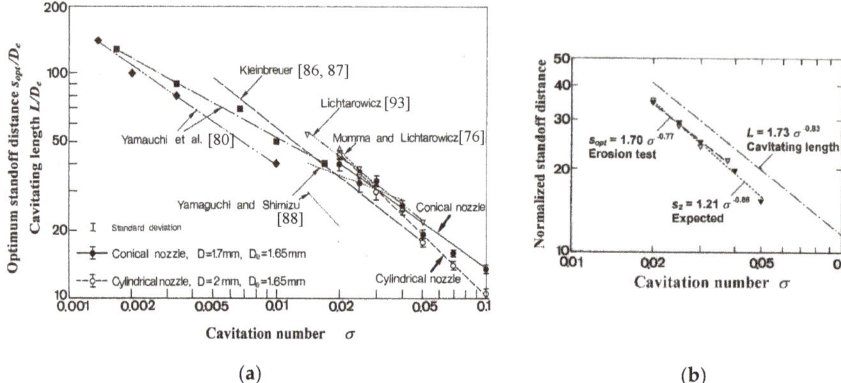

Figure 15. Relation between cavitation number and optimum standoff distance; (**a**) experimental result of optimum standoff distance and cavitating length; (**b**) estimated standoff distance from local cavitation number [72].

4.5. Sound Velocity in Cavitating Flow Field

In order to consider the key factors in cavitation impacts produced by the cavitating jet experimentally, sound velocity in the cavitating flow field has been investigated [95]. As erosion rate increases with acoustic impedance [96,97], the cavitation impact might increase with sound velocity in the cavitating flow field. In view of luminescence [62,98–100] and erosion [101], the effect of sound velocity of dissolved gas has been considered; however, there is no example of evaluating the sound velocity of the cavitating flow field itself. As is well known, the sound velocity changes drastically with the void ratio [102]. In the present review, the sound velocity in the cavitating flow field is considered.

To reveal the measuring method of the sound velocity of the cavitating flow field, Figure 16 shows the aspect of the cavitating flow through a Venturi tube. In Figure 16, the water flows from the left-hand side to the right-hand side. The cavitation occurs at the throat, and the vortex cavitation is observed at the end of the cavitating region. Further downstream of the vortex cavitation, many tiny bubbles are observed: they are residual bubbles [58] after cavitation collapses. As shown in Figure 16, the densities of residual bubbles differ greatly between the left and right sides of the white arrow. When this phenomenon is observed by a high-speed video camera as shown in Figure 17, the boundary of the density, which is marked by the yellow arrow in Figure 17, moves downstream. In the experiment, the upstream pressure in the absolute pressure of the throat was 0.6 MPa, and the ratio of the cross-sectional area between the throat and the tube was 9; thus, the flow speed at the downstream of the throat was about 3.5 m/s. On the other hand, the moving speed of the boundary marked by the yellow arrow was over 600 m/s. As shown in Figure 17, at the starting point of the boundary, the vortex cavitation shrank. This aspect suggests that the pressure wave is produced by vortex cavitation collapse, and the boundary movement reveals the pressure wave, as the residual bubbles are collapsed by the pressure wave. The sound velocity in the cavitating flow field can be estimated by measuring the moving speed of the pressure wave, i.e., the boundary of the density of the residual bubbles [95].

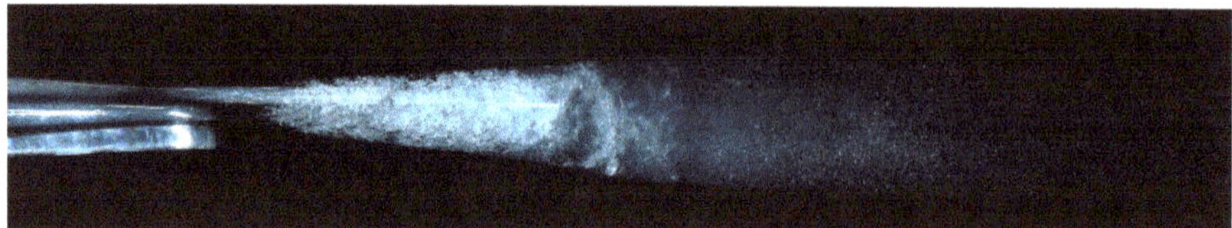

Figure 16. Aspects of vortex cavitation in the Venturi tube observed with a flash lamp. A pressure wave, which is indicated by the white arrow, was observed by collapsing residual bubbles.

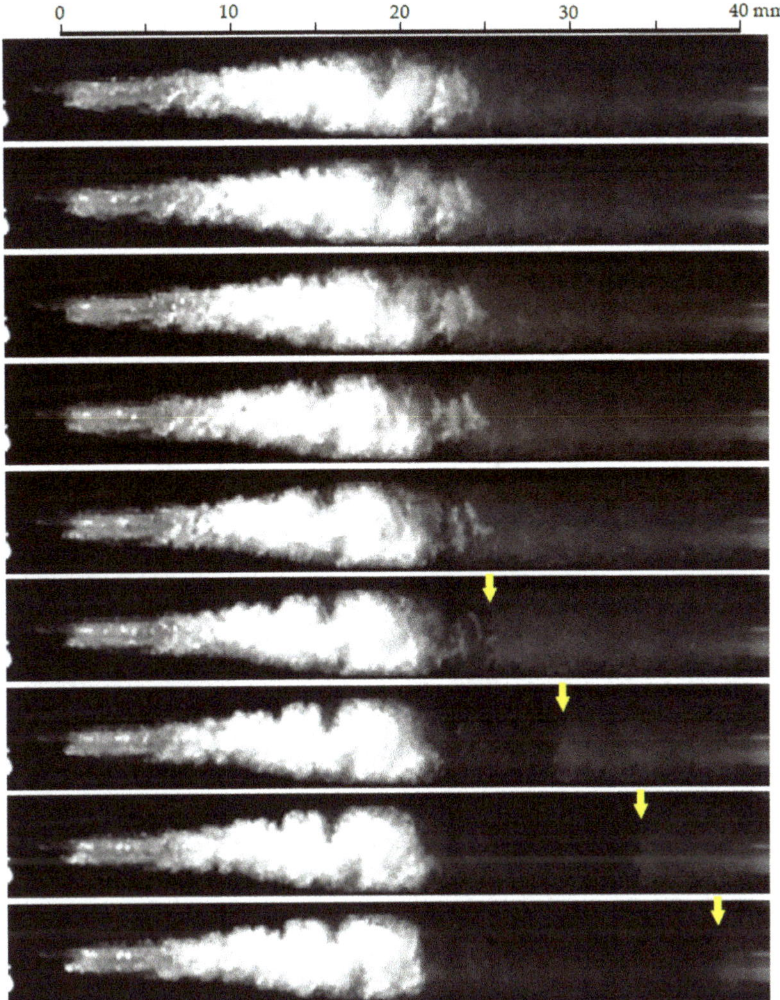

Figure 17. Aspects of vortex cavitation in Venturi tube observed by a high-speed video camera. The recording speed of the camera was 51,999 fps. A pressure wave, which is indicated by a yellow, arrow was observed after the vortex cavitation collapsed.

To explore the effect of cavitation number on the sound velocity, Figure 18 shows the aspect of the pressure wave changing with the cavitation number. The pressure wave is indicated by the yellow arrow in Figure 18. In Figure 18, the velocity of the pressure wave, i.e., the sound velocity, is shown in the right-hand side of the aspect. Figure 19 illustrates the relation between cavitation number and the sound velocity. The sound velocity v_s increases with cavitation number. At a relatively low void ratio, the sound velocity increases with a decrease in void ratio; thus, the tendency of the relation in Figure 18 is reasonable. As mentioned above, the erosion rate increased with the acoustic impedance [96,97], and the acoustic impedance is expressed as the product of the sound velocity, the density and the speed of the microjet; thus, the sound velocity in the cavitating flow field is a parameter of the cavitating jet. This result suggests that the increase in the sound velocity is one of reasons why the aggressive intensity of the cavitating jet increases with cavitation number, as the sound velocity increases with the cavitation number, as shown in Figure 19. On the other hand, the aggressive intensity of the cavitating jet decreases with an increase in cavitation number, as the cavitating region and/or bubble size decreases with an increase in the cavitation number. These two conflicting tendencies are the aggressive intensity of the cavitating jet has a peak for the cavitation number. More details are provided in "5. Estimation of Aggressive Intensity of Cavitating Jet". Note that the aggressive intensity of the cavitating jet had a peak at $\sigma = 0.01 - 0.014$ [4,103,104], and that of the hydrodynamic cavitation through Venturi tube had a peak at $\sigma = 0.4 - 0.7$ [99]. These peaks might depend on the flow pattern of the cavitating flow and/or the density of the residual bubbles.

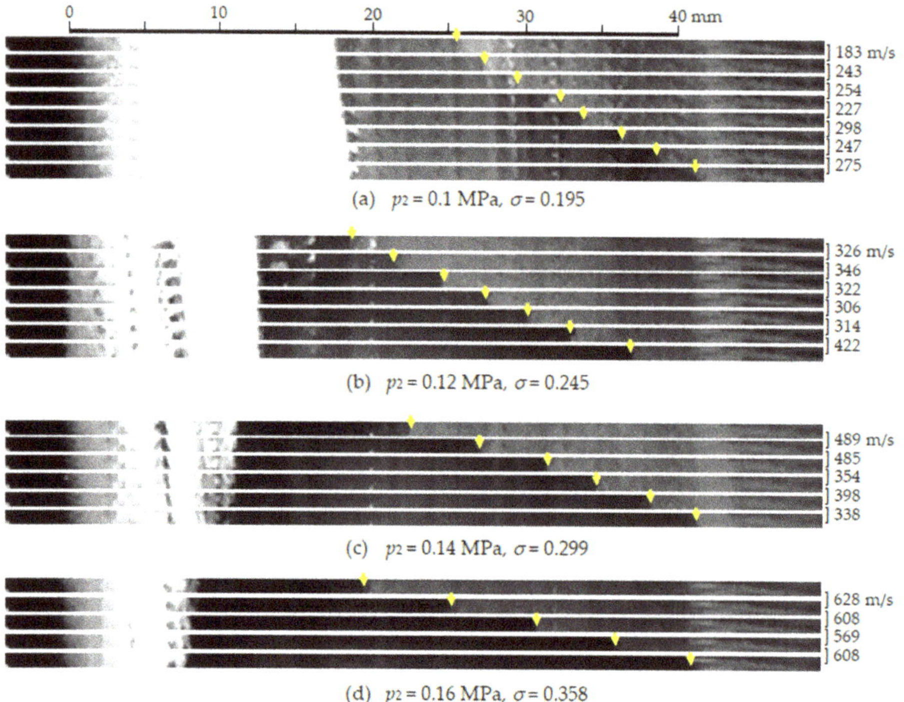

Figure 18. Aspects of pressure wave in Venturi tube changing with cavitation number. The recording speed of the camera was 109,999 fps.

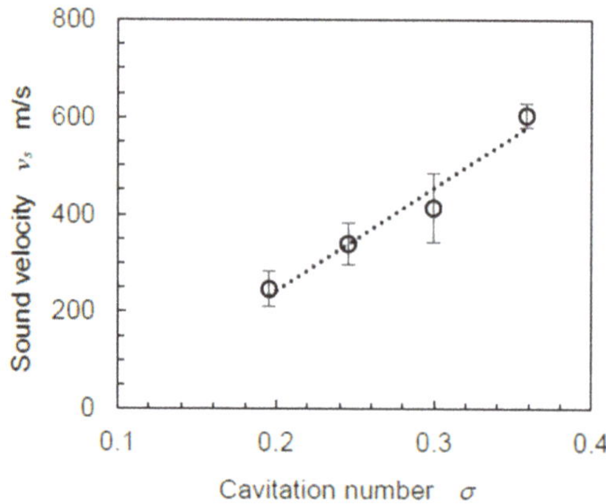

Figure 19. Sound velocity in cavitating flow field through Venturi tube changing with cavitation number.

4.6. Nozzle Geometry and Diameter

As mentioned above, the vortex cavitation in the cavitating jet is an important phenomenon, and the nozzle geometry, especially nozzle outlet geometry, is one of the key factors of the aggressive intensity of the cavitating jet, as the vortex is strongly affected by the nozzle outlet geometry. Figure 20 shows a schematic diagram of nozzle outlet geometry and the relative aggressive intensity of the cavitating jet from data published in previous reports [28,39,75,79]. Whereas a similar figure was introduced by Soyama [3], the data of nozzles K and L [28,75] were added in Figure 20. Nozzles A–E in Figure 20 are conventional nozzles for a water jet, and nozzle F is the standard nozzle for a standard test method for the erosion of solid materials by a cavitating jet [15]. Nozzle G is the nozzle obtained by optimizing the outlet bore and length experimentally [74]. Nozzle I and J had a guide pipe and a cavitator, as these enhance the aggressive intensity by about two times [39]. When both the guide pipe and the cavitator were installed, the aggressive intensity became four times larger that of nozzle J [39]. For nozzle K, when water flow holes were made near nozzle outlet, the aggressive intensity improved by 34%. When a long guide pipe with holes and water flow holes near the nozzle outlet were installed for nozzle L [28], the aggressive intensity of its cavitating jet L was 2.5 times larger than that of nozzle J and nearly 60 times larger than that of conventional water jet nozzles, as shown in Figure 20. The effect of nozzle geometry was also discussed in [14]. Resonating nozzles were also proposed and investigated [105,106].

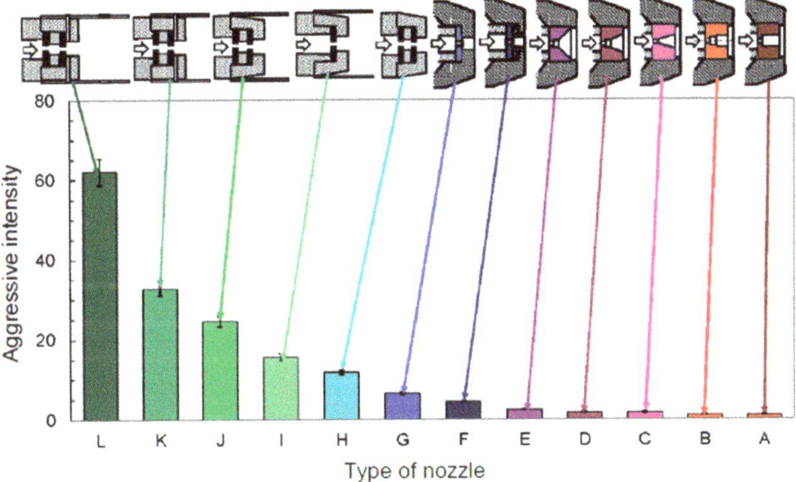

Figure 20. Effect of nozzle outlet geometry on aggressive intensity of cavitating jet.

The other important factor regarding nozzles is nozzle size. As mentioned above, vortex cavitation is an important phenomenon in the cavitating jet. In view of the Reynolds number, which is a key parameter of vortical flow, the larger velocity and the larger size are effective for the cavitating jet. However, too high a speed, i.e., too high an injection pressure, decreases the aggressive intensity of the cavitating jet, as shown in Figure 14b, as the sound velocity is decreased at too low a cavitation number, i.e., too-high speed condition. Thus, in the case of practical applications of the cavitating jet, the cavitating jet using a large nozzle at a relatively low injection pressure is better than that of a small nozzle at a high injection pressure [80]. The scaling law of nozzle size on the aggressive intensity of the cavitating jet is discussed in reference [107].

4.7. Water Qualities

As cavitation is a phase-change phenomenon from liquid phase to gas phase, as mentioned above, the water temperature affects the aggressive intensity of the cavitating jet [108–110]. It was reported that peening intensity using the cavitating jet at 288–308 K was nearly constant [4]. Whereas cavitation nuclei are required for the cavitating jet, too many air bubbles reduce the aggressive intensity of the cavitating jet due to the cushion effect [2] and the decrease in the sound velocity. When the water-filled chamber for the cavitating jet is too small, the suction vortex caused by the jet reduces the aggressive intensity of the cavitating jet. Thus, in the report, the effects of water depth and the chamber size on the aggressive intensity were investigated experimentally [111]. In the report [111], the effect of gas content on the peening intensity using the cavitating jet was also investigated using degassed water.

5. Estimation of Aggressive Intensity of Cavitating Jet

As mentioned above, in the constant downstream pressure condition, too high an injection pressure, i.e., too low a cavitation number, reduces the aggressive intensity of the cavitating jet. In this section, the mechanism is discussed, and a method to estimate the aggressive intensity of the cavitation jet as a function of cavitation number is proposed.

In Figure 21, the experimental results of the aggressive intensity of the cavitating jet as a function of cavitation is revealed by blue closed circles [112]. The aggressive intensity is normalized by the maximum value, and it has a peak at $\sigma = 0.016$. As the energy of cavitation is proportional to the volume of the cavitation and the pressure difference of the bubble [111], the aggressive intensity of the cavitating jet I_{cav} can be assumed as follows.

$$I_{cav} \propto (L_{cav})^3 \cdot (p_2 - p_v) \tag{13}$$

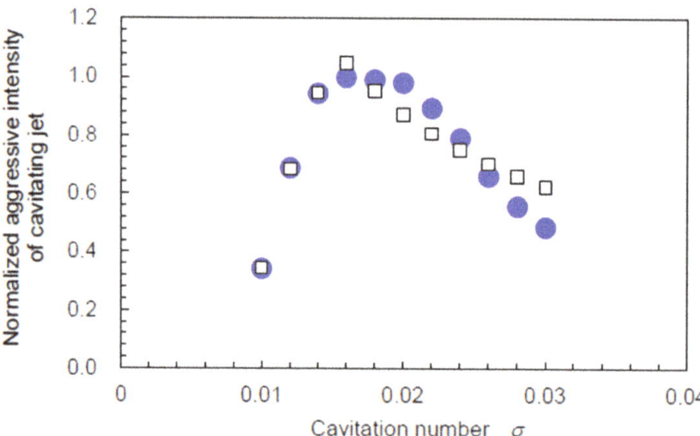

Figure 21. Aggressive intensity of cavitating jet changing with cavitation number.

As I_{cav} is affected by the acoustic impedance [95,111], the term of the sound velocity v_s is included in Equation (13). As I_{cav} is also affected by flow velocity, which is defined by the pressure difference, i.e., $\sqrt{p_1 - p_2}$, the velocity term is also included in Equation (13).

$$I_{cav} \propto (L_{cav})^3 \cdot (p_2 - p_v) \cdot (v_s - v_{s\,th}) \cdot \left(\sqrt{p_1 - p_2}\right)^n \tag{14}$$

Here, $v_{s\,th}$ is the threshold level of the sound velocity considering the threshold level of I_{cav} [43]. The n is put as an exponent in Equation (14) to consider the power law of the velocity on I_{cav} [76,105,111–113].

If it was assumed that I_{cav} was proportional to the flow energy, which was defined by the product of the pressure and the flow rate, the flow energy is proportional to $p_1 - p_2$ and $\sqrt{p_1 - p_2}$, as $\sqrt{p_1 - p_2}$ was proportional to the flow velocity. Namely, the flow energy is proportional to cube of $\sqrt{p_1 - p_2}$. Thus, in the present review, n = 3 is chosen. From Equations (8) and (14), Equation (15) was obtained.

$$I_{cav} = c_3\, \sigma^{-1.8} \cdot (p_2 - p_v) \cdot (v_s - v_{s\ th}) \cdot \left(\sqrt{p_1 - p_2}\right)^3 \tag{15}$$

Here, c_3 is constant. When Equation (15) is assumed, I_{cav} can be estimated by obtaining c_3, c_4, c_5 and $v_{s\ th}$ by a least-square method.

$$v_s = c_4\, \sigma + c_5 \tag{16}$$

The estimated I_{cav} is shown in Figure 21 by empty squares. The correlation coefficient between experimental data and estimated values is 0.924. As the number of the datasets was 11, the probability of non-correlation is less than 0.005%. If the probability of a non-correlation is less than 1%, it can be concluded that the relationship is highly significant. Thus, it can be concluded that the relationship between the experimental data and the estimated values is highly significant. I_{cav} can be estimated by Equation (15).

6. Applications of Cavitating Jet

The cavitating jet has been applied for drilling and cutting rocks [16,105,106,113]. The cavitating jet in air can also dig concrete structures for the maintenance of infrastructure [114]. In the case of mechanical surface treatment, the use of a submerged water jet was attempted to mitigate stress corrosion cracking (SCC) of nuclear power plants by using impinging impacts of a water column in the jet center at the beginning [84]. Soyama et al. found that the cavitating jet could introduce compressive residual stress into stainless steel by using cavitation impacts [13]; then, the cavitating jet was successfully applied to mitigate SCC in nuclear power plants [115]. Based on experimental results of the introduction of compressive residual stress into metallic materials, the improvement in the fatigue strength of metallic components by cavitation peening was proposed using a pressurized chamber to enhance the aggressive intensity of the cavitating jet [116–118], and it was demonstrated for forging die [119], gears [120,121], continuous valuable transmission CVT elements [122] and rollers [123]. After enhancing the aggressive intensity of the cavitating jet by optimizing the nozzle geometry, the improvement in fatigue strength by cavitation peening with an open chamber was demonstrated [17,124–127]. Cavitation peening using oil was also proposed [128]. Cavitation peening also improves tribological properties such as fretting fatigue properties [129,130]. The improvement in fatigue strength of metallic materials using a cavitating jet in air was also demonstrated [36,131,132]. In view of environmental-assisted cracking, the suppression of hydrogen-assisted fatigue crack growth in austenitic stainless steel and delayed fracture resistance on chrome molybdenum steel by cavitation peening were reported [133,134].

The cavitating jet can be applied in the semiconductor industry. It can be used not only for cleaning [135], but also gettering [136]. When erosion rates using ultrasonic vibratory apparatus [137] and cavitating jet apparatus [15] were compared, the erosion rate of the cavitating jet apparatus was larger than that of ultrasonic vibratory apparatus. As conventional ultrasonic cleaning devices use ultrasonic cavitation, cleaning using the cavitating jet is more powerful. The gettering technique is a method to remove unwanted impurities from active device regions in a silicon wafer into the back side of the wafer by introducing oxidation-induced stacking faults (OSF) [138–142]. In order to produce OSF, the introduction of strain into the wafer is required, and shot blasting is used in a conventional way [143,144]. However, the cleaning of shots is required. In the case of gettering using the cavitating jet, the cavitating jet can introduce strain for OSF [145,146] and cleaning at the same time. This is a great advantage for semiconductor processes.

One of the applications that was proposed in the bioengineering area is oral cleaning using the cavitating jet [147–149]. Dental implants have been used as the solution for the loss of teeth;

however, peri-implant mucosis and peri-implantitis are new dental diseases affecting implants [150,151]. The most effective treatment for these diseases is cleaning dental plaque, which is a kind of biofilm that adheres to the surface of teeth and implants [152]. Conventional cleaning methods of the dental plaque are oral brushing, ultrasonic scaling and rubber cup cleaning. Unfortunately, micro-textured roughness is made on the implant surface to improve biocompatibility [153], so oral toothbrushes and the tips of scaling instruments cannot reach the bottom of the rough surface. The cleaning of dental plaque on the rough surface of dental implants using the cavitating jet has been successfully demonstrated [147–149].

As is well known, sonochemistry is a research area of chemistry using ultrasonic cavitation [154]. Hydrodynamic cavitation such as the cavitating jet and cavitating flow through orifices can be applied for wastewater treatment [155–162] and to oxidize organic compounds [163]. The dispersion of spilled oil by a cavitating jet at sea has also been proposed [164]. It was reported that the efficacy of hydrodynamic cavitation was 20 times better than that of ultrasonic cavitation when the efficiency of the hydrodynamic cavitation on the pretreatment of biomass was compared with that of ultrasonic cavitation [165].

7. Conclusions

To use the cavitating jet for practical applications, the unsteady behavior of the cavitating jet, i.e., a submerged water jet with cavitation, was reviewed. The key factors on the aggressive intensity of the cavitating jet were also summarized. In the present review, the aggressive intensity of the cavitating jet was investigated by erosion rate and/or peening intensity. The main topics reviewed in the paper are summarized as follows.

(1) The cavitation is initiated inside and/or outside of the nozzle as a ring or helical vortex cavitation. These vortex cavitations become cloud cavitations combining with each other, and the cloud cavitation sheds periodically;
(2) Cloud shedding is a phenomenon governed by a constant Strouhal number, which is defined by the shedding frequency, the flow velocity and the width of the cavitating region;
(3) The cloud cavitation forms a ring vortex cavitation on the impinging surface and then collapses producing impacts;
(4) At optimum conditions, the affected area on the flat target by the impinging cavitating jet is a ring. The mechanism of the ring region can be explained by considering the local cavitation number on the surface. Note that the ring does not directly result from the swirl around the cavitating jet;
(5) At constant downstream pressure conditions, the aggressive intensity of the cavitating jet increases with the injection pressure, and it saturates at a certain pressure and then decreases. At too high an injection pressure, the aggressive intensity decreases.;
(6) At constant injection pressure conditions, the aggressive intensity of the cavitating jet increases with a decrease in cavitation number σ, and it saturates at $\sigma = 0.01 - 0.02$ and then decreases at too low σ;
(7) One reason why the aggressive intensity of the cavitating jet decreases at too high an injection pressure, i.e., too low a cavitation number, is the decrease in sound velocity in the cavitating flow field.

Funding: This research was partly supported by JSPS KAKENHI grant numbers 18KK0103 and 20H02021.

Conflicts of Interest: The author declares no conflict of interest.

References

1. Soyama, H.; Oba, R.; Kato, H. Cavitation Observations of Severely Erosive Vortex Cavitation Arising in a Centrifugal Pump. In Proceedings of the Institution of Mechanical Engineers, 3rd International Conference on Cavitation, Cambrige, UK, 9–11 December 1992; pp. 103–110.
2. Brennen, C.E. *Cavitation and Bubble Dynamics*; Oxford University Press: New York, NY, USA, 1995.
3. Soyama, H. Cavitation peening: A review. *Metals* **2020**, *10*, 270. [CrossRef]
4. Soyama, H. Key factors and applications of cavitation peening. *Inter. J. Peen. Sci. Technol.* **2017**, *1*, 3–60.
5. Soyama, H.; Saito, K.; Saka, M. Improvement of fatigue strength of aluminum alloy by cavitation shotless peening. *J. Eng. Mater. Technol.* **2002**, *124*, 135–139. [CrossRef]
6. Soyama, H.; Chighizola, C.R.; Hill, M.R. Effect of compressive residual stress introduced by cavitation peening and shot peening on the improvement of fatigue strength of stainless steel. *J. Mater. Process. Technol.* **2021**, *288*, 1168771–11687710, paper No. 116877. [CrossRef]
7. Mankbadi, R.R. Dynamics and control of coherent structure in turbulent jets. *Appl. Mech. Rev.* **1992**, *45*, 219–248. [CrossRef]
8. Taghavi, R. Cavitation Inception in Axisymmertric Jets. Ph.D. Thesis, University of Minnesota, Minneapolis, MN, USA, 1985.
9. Soyama, H.; Lichtarowicz, A.; Momma, T. Vortex Cavitation in a Submerged Jet. In Proceedings of the 1996 ASME Fluids Engineering Division Summer Meeting, San Diego, CA, USA, 7–11 July 1996; pp. 415–422, FED-236.
10. Arndt, R.E.A. Cavitation in vortical flows. *Annu. Rev. Fluid Mech.* **2002**, *34*, 143–175. [CrossRef]
11. Lichtarowicz, A. Use of a simple cavitating nozzle for cavitation erosion testing and cutting. *Nat. Phys. Sci.* **1972**, *239*, 63–64. [CrossRef]
12. Yamauchi, Y.; Soyama, H.; Adachi, Y.; Sato, K.; Shindo, T.; Oba, R.; Oshima, R.; Yamabe, M. Suitable region of high-speed submerged water jets for cutting and peening. *JSME Int. J.* **1995**, *38B*, 31–38. [CrossRef]
13. Soyama, H.; Yamauchi, Y.; Ikohagi, T.; Oba, R.; Sato, K.; Shindo, T.; Oshima, R. Marked peening effects by highspeed submerged-water-jets—Residual stress change on SUS304. *Jet Flow Eng.* **1996**, *13*, 25–32.
14. Hutli, E.; Nedeljkovic, M.S.; Bonyar, A.; Legrady, D. Experimental study on the influence of geometrical parameters on the cavitation erosion characteristics of high speed submerged jets. *Exp. Fluid Sci.* **2017**, *80*, 281–292. [CrossRef]
15. G134-17 Standard test method for erosion of solid materials by a cavitating liquid jet. *ASTM Stand.* **2017**, *03.02*, 1–17.
16. Peng, C.; Tian, S.C.; Li, G.S. Joint experiments of cavitation jet: High-speed visualization and erosion test. *Ocean Eng.* **2018**, *149*, 1–13. [CrossRef]
17. Soyama, H. Comparison between the improvements made to the fatigue strength of stainless steel by cavitation peening, water jet peening, shot peening and laser peening. *J. Mater. Process. Technol.* **2019**, *269*, 65–78. [CrossRef]
18. Soyama, H.; Yamauchi, Y.; Adachi, Y.; Sato, K.; Shindo, T.; Oba, R. High-speed observations of the cavitation cloud around a high-speed submerged water-jet. *JSME Int. J.* **1995**, *38B*, 245–251. [CrossRef]
19. Hutli, E.A.F.; Nedeljkovic, M.S. Frequency in shedding/discharging cavitation clouds determined by visualization of a submerged cavitating jet. *J. Fluids Eng.* **2008**, *130*, 1–8, paper No. 021304. [CrossRef]
20. Sato, K.; Sugimoto, Y.; Ohjimi, S. Structure of periodic cavitation clouds in submerged impinging water-jet issued from horn-type nozzle. In Proceedings of the 9th Pacific Rim International Conference on Water Jetting Technology, Koriyama, Japan, 20–23 November 2009.
21. Nishimura, S.; Takakuwa, O.; Soyama, H. Similarity law on shedding frequency of cavitation cloud induced by a cavitating jet. *J. Fluid Sci. Technol.* **2012**, *7*, 405–420. [CrossRef]
22. Wright, M.M.; Epps, B.; Dropkin, A.; Truscott, T.T. Cavitation of a submerged jet. *Exp. Fluids* **2013**, *54*, 21. [CrossRef]
23. Sato, K.; Taguchi, Y.; Hayashi, S. High speed observation of periodic cavity behavior in a convergent-divergent nozzle for cavitating water jet. *J. Flow Control Meas. Vis.* **2013**, *1*, 102–107. [CrossRef]
24. Liu, H.X.; Kang, C.; Zhang, W.; Zhang, T. Flow structures and cavitation in submerged waterjet at high jet pressure. *Exp. Fluid Sci.* **2017**, *88*, 504–512. [CrossRef]
25. Hutli, E.; Nedeljkovic, M.; Bonyar, A. Dynamic behaviour of cavitation clouds: Visualization and statistical analysis. *J. Braz. Soc. Mech. Sci. Eng.* **2019**, *41*, 15. [CrossRef]

26. Wu, Q.; Wei, W.; Deng, B.; Jiang, P.; Li, D.; Zhang, M.D.; Fang, Z.L. Dynamic characteristics of the cavitation clouds of submerged helmholtz self-sustained oscillation jets from high-speed photography. *J. Mech. Sci. Technol.* **2019**, *33*, 621–630. [CrossRef]
27. Liu, B.; Pan, Y.; Ma, F. Pulse pressure loading and erosion pattern of cavitating jet. *Eng. Appl. Comp. Fluid Mech.* **2020**, *14*, 136–150. [CrossRef]
28. Kamisaka, H.; Soyama, H. Enhancing the aggressive intensity of a cavitating jet by introducing water flow holes and a long guide pipe. *J. Fluids Eng.Trans. ASME* **2020**, in press. [CrossRef]
29. Gopalan, S.; Katz, J.; Knio, O. The flow structure in the near field of jets and its effect on cavitation inception. *J. Fluid Mech.* **1999**, *398*, 1–43. [CrossRef]
30. Arndt, R.E.A.; Amromin, E.L.; Hambleton, W. Cavitation inception in the wake of a jet-driven body. *J. Fluids Eng.Trans. ASME* **2009**, *131*, 8. [CrossRef]
31. Brandner, P.A.; Pearce, B.W.; de Graaf, K.L. Cavitation about a jet in crossflow. *J. Fluid Mech.* **2015**, *768*, 34. [CrossRef]
32. Yuan, C.; Song, J.C.; Liu, M.H. Comparison of compressible and incompressible numerical methods in simulation of a cavitating jet through a poppet valve. *Eng. Appl. Comp. Fluid Mech.* **2019**, *13*, 67–90. [CrossRef]
33. Soyama, H. Introduction of compressive residual stress using a cavitating jet in air. *J. Eng. Mater. Technol.* **2004**, *126*, 123–128. [CrossRef]
34. Soyama, H.; Kikuchi, T.; Nishikawa, M.; Takakuwa, O. Introduction of compressive residual stress into stainless steel by employing a cavitating jet in air. *Surf. Coat. Technol.* **2011**, *205*, 3167–3174. [CrossRef]
35. Soyama, H. High-speed observation of a cavitating jet in air. *J. Fluids Eng.* **2005**, *127*, 1095–1108. [CrossRef]
36. Soyama, H. Improvement of fatigue strength by using cavitating jets in air and water. *J. Mater. Sci.* **2007**, *42*, 6638–6641. [CrossRef]
37. Marcon, A.; Melkote, S.N.; Yoda, M. Effect of nozzle size scaling in co-flow water cavitation jet peening. *J. Manuf. Process.* **2018**, *31*, 372–381. [CrossRef]
38. Marcon, A.; Melkote, S.N.; Yoda, M.; Sanders, D. Analysis of co-flow water cavitation peening of Al7075-T651 alloy using high-speed imaging and surface pitting tests. *Mater. Perform. Charact.* **2018**, *7*, 1018–1040. [CrossRef]
39. Soyama, H. Enhancing the aggressive intensity of a cavitating jet by introducing a cavitator and a guide pipe. *J. Fluid Sci. Technol.* **2014**, *9*, 1–12, paper No. 13-00238. [CrossRef]
40. Kamisaka, H.; Soyama, H. Periodical shedding of cavitation cloud induced by a cavitating jet. In Proceedings of the 24th International Conference on Water Jetting, Manchester, UK, 5–6 September 2018; 2018; pp. 111–123.
41. Soyama, H.; Lichtarowicz, A.; Momma, T.; Williams, E.J. A new calibration method for dynamically loaded transducers and its application to cavitation impact measurement. *J. Fluids Eng.* **1998**, *120*, 712–718. [CrossRef]
42. Soyama, H.; Sekine, Y.; Saito, K. Evaluation of the enhanced cavitation impact energy using a PVDF transducer with an acrylic resin backing. *Measurement* **2011**, *44*, 1279–1283. [CrossRef]
43. Soyama, H.; Kumano, H. The fundamental threshold level—A new parameter for predicting cavitation erosion resistance. *J. Test. Eval.* **2002**, *30*, 421–431.
44. Schnerr, G.H.; Sezal, I.H.; Schmidt, S.J. Numerical investigation of three-dimensional cloud cavitation with special emphasis on collapse induced shock dynamics. *Phys. Fluids* **2008**, *20*, 9. [CrossRef]
45. Bensow, R.E.; Bark, G. Implicit les predictions of the cavitating flow on a propeller. *J. Fluids Eng.Trans. ASME* **2010**, *132*, 10. [CrossRef]
46. Ji, B.; Luo, X.W.; Wu, Y.L.; Peng, X.X.; Duan, Y.L. Numerical analysis of unsteady cavitating turbulent flow and shedding horse-shoe vortex structure around a twisted hydrofoil. *Int. J. Multiph. Flow* **2013**, *51*, 33–43. [CrossRef]
47. Xing, T.; Li, Z.Y.; Frankel, S.H. Numerical simulation of vortex cavitation in a three-dimensional submerged transitional jet. *J. Fluids Eng.Trans. ASME* **2005**, *127*, 714–725. [CrossRef]
48. Alehossein, H.; Qin, Z. Numerical analysis of rayleigh-plesset equation for cavitating water jets. *Int. J. Numer. Methods Eng.* **2007**, *72*, 780–807. [CrossRef]
49. Sonde, E.; Chaise, T.; Boisson, N.; Nelias, D. Modeling of cavitation peening: Jet, bubble growth and collapse, micro-jet and residual stresses. *J. Mater. Process. Technol.* **2018**, *262*, 479–491. [CrossRef]

50. Dittakavi, N.; Chunekar, A.; Frankel, S. Large eddy simulation of turbulent-cavitation interactions in a venturi nozzle. *J. Fluids Eng.Trans. ASME* **2010**, *132*, 11. [CrossRef]
51. Peng, G.Y.; Shimizu, S. Progress in numerical simulation of cavitating water jets. *J. Hydrodyn.* **2013**, *25*, 502–509. [CrossRef]
52. Peng, G.Y.; Yang, C.X.; Oguma, Y.; Shimizui, S. Numerical analysis of cavitation cloud shedding in a submerged water jet. *J. Hydrodyn.* **2016**, *28*, 986–993. [CrossRef]
53. Hsiao, C.T.; Jayaprakash, A.; Kapahi, A.; Choi, J.K.; Chahine, G.L. Modelling of material pitting from cavitation bubble collapse. *J. Fluid Mech.* **2014**, *755*, 142–175. [CrossRef]
54. Choi, J.K.; Chahine, G.L. Relationship between material pitting and cavitation field impulsive pressures. *Wear* **2016**, *352–353*, 42–53. [CrossRef]
55. Ma, J.S.; Hsiao, C.T.; Chahine, G.L. Numerical study of acoustically driven bubble cloud dynamics near a rigid wall. *Ultrason. Sonochem.* **2018**, *40*, 944–954. [CrossRef]
56. Sasaki, H.; Iga, Y.; Soyama, H. Effect of Bubble Radius on Ability of Submerged Laser Peening. In *Advanced Surface Enhancement, INCASE 2019, Lecture Notes in Mechanical Engineering*; Springer Nature Singapore Pte Ltd.: Singapore, 2020; pp. 283–291.
57. Lauterborn, W.; Bolle, H. Experimental investigations of cavitation-bubble collapse in neighborhood of a solid boundary. *J. Fluid Mech.* **1975**, *72*, 391–399. [CrossRef]
58. Soyama, H. Corrosion behavior of pressure vessel steel exposed to residual bubbles after cavitation bubble collapse. *Corrosion* **2011**, *67*, 1–8, paper No. 025001. [CrossRef]
59. Plesset, M.S.; Chapman, R.B. Collapse of an initially spherical vapour cavity in neighbourhood of a solid boundary. *J. Fluid Mech.* **1971**, *47*, 283–290. [CrossRef]
60. Crum, L.A. Surface oscillations and jet development in pulsating bubbles. *J. Phys. Colloq.* **1979**, *40*, 285–288. [CrossRef]
61. Lauterborn, W.; Ohl, C.D. Cavitation bubble dynamics. *Ultrason. Sonochem.* **1997**, *4*, 65–75. [CrossRef]
62. Lauterborn, W.; Kurz, T. Physics of bubble oscillations. *Rep. Prog. Phys.* **2010**, *73*, 88. [CrossRef]
63. Soyama, H.; Ohba, K.; Takeda, S.; Oba, R. High-speed observations of highly erosive vortex cavitation around butterfly valve. *Trans. JSME* **1994**, *60B*, 1133–1138. [CrossRef]
64. Ohba, K.; Soyama, H.; Takeda, S.; Inooka, H.; Oba, R. High-speed observations of highly erosive vortex cavitation using image processing. *J. Flow Vis. Image Process.* **1995**, *2*, 161–172. [CrossRef]
65. Soyama, H.; Li, S.R.; Tonosaki, M.; Uranishi, K.; Kato, H.; Oba, R. High-speed observations of severe cavitation erosion in a high-specific-speed centrifugal pump. *Trans. JSME* **1995**, *61B*, 3945–3951. [CrossRef]
66. Dominguez Cortazar, M.A.; Franc, J.P.; Michel, J.M. The erosive axial collapse of a cavitating vortex: An experimental study. *J. Fluids Eng.* **1997**, *119*, 686–691. [CrossRef]
67. Takakuwa, O.; Soyama, H. The effect of scanning pitch of nozzle for a cavitating jet during overlapping peening treatment. *Surf. Coat. Technol.* **2012**, *206*, 4756–4762. [CrossRef]
68. Ohya, T.; Okimura, K.; Ohta, T.; Ichioka, T. Residual Stress Improved by Water Jet Peening for Small-Diameter Pipe Inner Surfaces. In *Mitsubishi Heavy Industries, Ltd. Technical Review*; Mitsubishi: Tokyo, Japan, 2000; Volume 37, pp. 52–55.
69. Demma, A.; Frederick, G. Program on Technology Innovation: An Evaluation of Surface Stress Improvement Technologies for Pwscc Mitigation of Alloy 600 Nuclear Components. In *Materials Reliability Program (MRP-162)*; Electric Power Research Institute: Palo Alto, CA, USA, 2006; pp. 1–104.
70. Crooker, P.; Lian, T. *Materials Reliability Program: Technical Basis for Primary Water Stress Corrosion Cracking Mitigation by Surface Stress Improvement (MRP-267, Revision 1)*; Electric Research Power Institute: Palo Alto, CA, USA, 2012; pp. 1–330.
71. Hutli, E.; Nedeljkovic, M.S.; Radovic, N.A.; Bonyar, A. The relation between the high speed submerged cavitating jet behaviour and the cavitation erosion process. *Int. J. Multiph. Flow* **2016**, *83*, 27–38. [CrossRef]
72. Soyama, H.; Lichtarowicz, A. Cavitating jets—Similarity correlations. *J. Jet Flow Eng.* **1996**, *13*, 9–19.
73. Nobel, A.J.; Talmon, A.M. Measurements of the stagnation pressure in the center of a cavitating jet. *Exp. Fluids* **2012**, *52*, 403–415. [CrossRef]
74. Soyama, H. Enhancing the aggressive intensity of a cavitating jet by means of the nozzle outlet geometry. *J. Fluids Eng.* **2011**, *133*, 1–11, paper No. 101301. [CrossRef]
75. Kamisaka, H.; Soyama, H. Enhancement of an aggressive intensity of a cavitating jet by water flow holes near nozzle outlet. *Trans. JSME* **2019**, *85*. paper No. 19-00280. [CrossRef]

76. Soyama, H. Fundamentals and Applications of Cavitation Peening Comparing with Shot Peening and Laser Peening. In *Advanced Surface Enhancement, INCASE 2019, Lecture Notes in Mechanical Engineering*; Springer Nature Singapore Pte Ltd.: Singapore, 2020; pp. 76–87.
77. Momma, T.; Lichtarowicz, A. A study of pressure and erosion produced by collapsing cavitation. *Wear* **1995**, *186*, 425–436. [CrossRef]
78. Soyama, H.; Nagasaka, K.; Takakuwa, O.; Naito, A. Optimum injection pressure of a cavitating jet for introducing compressive residual stress into stainless steel. *J. Power Energy Syst.* **2012**, *6*, 63–75. [CrossRef]
79. Soyama, H. Effect of nozzle geometry on a standard cavitation erosion test using a cavitating jet. *Wear* **2013**, *297*, 895–902. [CrossRef]
80. Soyama, H.; Takakuwa, O. Enhancing the aggressive strength of a cavitating jet and its practical application. *J. Fluid Sci. Technol.* **2011**, *6*, 510–521. [CrossRef]
81. Yamauchi, Y.; Soyama, H.; Sato, K.; Ikohagi, T.; Oba, R. Development of erosion in high-speed submerged water jets. *Trans. JSME* **1994**, *60B*, 736–743. [CrossRef]
82. Yamauchi, Y.; Asami, K.; Soyama, H.; Sato, K.; Ikohagi, T.; Oba, R. Marked effects of ambient pressure on impinging erosion induced by high-speed submerged water jets. *Trans. JSME* **1995**, *61B*, 785–792. [CrossRef]
83. Hirano, K.; Enomoto, K.; Hayashi, E.; Kurosawa, K. Effect of water jet peening on corrosion resistance and fatigue strength of type 304 stainless steel. *J. Soc. Mater. Soc. Jpn.* **1996**, *45*, 740–745. [CrossRef]
84. Enomoto, K.; Hirano, K.; Mochizuki, M.; Kurosawa, K.; Saito, H.; Hayashi, E. Improvement of residual stress on material surface by water jet peening. *J. Soc. Mater. Soc. Jpn.* **1996**, *45*, 734–739. [CrossRef]
85. Lichtarowicz, A.; Scott, P.J. Erosion Testing with Cavitating Jet. In *Proceedings of 5th International Conference on Erosion by Liquid and Solid Impact*; Office of Naval Research London: London, UK, 1979; paper No. 69.
86. Kleinbreuer, W. Werkstoffzerstorung durch kavitation in olhydralishen systemen. *Ind. Anz.* **1976**, *98*, 1096–1100.
87. Kleinbreuer, W. Kavitationserosion in hydraulischen systemen. *Ind. Anz.* **1977**, *99*, 609–613.
88. Lichtarowicz, A. Cavitating jet apparatus for cavitation erosion testing, erosion: Prevention and useful applications. *ASTM STP* **1979**, *669*, 530–549.
89. Yamaguchi, A.; Shimizu, S. Erosion due to impingement of cavitating jet. *J. Fluids Eng.Trans. ASME* **1987**, *109*, 442–447. [CrossRef]
90. Soyama, H. Surface mechanics design of metallic materials on mechanical surface treatments. *Mech. Eng. Rev.* **2015**, *2*, 1–20, paper No. 14-00192. [CrossRef]
91. Soyama, H.; Sanders, D. Use of an abrasive water cavitating jet and peening process to improve the fatigue strength of titanium alloy 6Al-4V manufactured by the electron beam powder bed melting (EBPB) additive manufacturing method. *JOM* **2019**, *71*, 4311–4318. [CrossRef]
92. Kamisaka, H.; Soyama, H. Effect of injection pressure on mechanical surface treatment using a submerged water jet. *J. Jet Flow Eng.* **2018**, *33*, 4–10.
93. J443 Procedures for Using Standard Shot Peening Almen Strip. In *SAE International Standards*; SAE International: Warrendale, PA, USA, 2010; pp. 1–6.
94. Lichtarowicz, A. Erosion testing with cavitating jet. In Proceedings of the Cavitation Erosion in Fluid Systems, ASME Fluids Engineering Conference, Boulder, CO, USA, 22–24 June 1981; pp. 153–161.
95. Soyama, H. Estimation of luminescence intensity of hydrodynamic cavitation considering sound velocity in the cavitating flow field. In Proceedings of the 11th International Symposium on Cavitation, Daejeon, Korea, 9–13 May 2021.
96. Wilson, R.W.; Graham, R. Cavitation of metal surfaces in contact with lubricants. *Conf. Lubr. Wear IME* **1957**, 707–712.
97. Hattori, S.; Motoi, Y.; Kikuta, K.; Tomaru, H. Cavitation Erosion Of Silver Plated Coating at Different Temperatures and Pressures. In *8th International Symposium on Measurement Techniques for Multiphase Flows*; Li, Y., Zheng, Y., Eds.; American Institute of Physics: Melville, NY, USA, 2014; pp. 358–365.
98. Bass, A.; Ruuth, S.J.; Camara, C.; Merriman, B.; Putterman, S. Molecular dynamics of extreme mass segregation in a rapidly collapsing bubble. *Phys. Rev. Lett.* **2008**, *101*, 4. [CrossRef] [PubMed]
99. Soyama, H.; Hoshino, J. Enhancing the aggressive intensity of hydrodynamic cavitation through a venturi tube by increasing the pressure in the region where the bubbles collapse. *AIP Adv.* **2016**, *6*, 1–13, paper No. 045113. [CrossRef]
100. Gielen, B.; Marchal, S.; Jordens, J.; Thomassen, L.C.J.; Braeken, L.; Van Gerven, T. Influence of dissolved gases on sonochemistry and sonoluminescence in a flow reactor. *Ultrason. Sonochem.* **2016**, *31*, 463–472. [CrossRef]

101. Kang, B.K.; Kim, M.S.; Park, J.G. Effect of dissolved gases in water on acoustic cavitation and bubble growth rate in 0.83 MHz megasonic of interest to wafer cleaning. *Ultrason. Sonochem.* **2014**, *21*, 1496–1503. [CrossRef]
102. Kieffer, S.W. Sound speed in liquid-gas mixtures—Water-air and water-steam. *J. Geophys. Res.* **1977**, *82*, 2895–2904. [CrossRef]
103. Shimizu, S.; Tanioka, K.; Ikegami, N. Erosion due to ultra-high-speed cavitating jet. *J. Jpn. Hydraul. Pneum. Soc.* **1997**, *28*, 778–784.
104. Soyama, H. Material testing and surface modification by using cavitating jet. *J. Soc. Mater. Sci. Jpn.* **1998**, *47*, 381–387. [CrossRef]
105. Johnson, V.E.; Chahine, G.L.; Lindenmuth, W.T.; Conn, A.F.; Frederick, G.S.; Giacchino, G.J. Cavitating and structured jets for mechanical bits to increase drilling rate, 1. Tehory and concepts. *J. Energy Resour. Technol.* **1984**, *106*, 282–288. [CrossRef]
106. Chahine, G.L.; Johnson, V.E., Jr.; Kalumuck, K.M.; Perdue, T.O.; Waxman, D.N.; Frederick, G.S.; Watson, R.E. Internal and external acoustics and large structure dynamics of cavitating self-resonating water jets. In *Sandia National Laboratories, Contractor Report*; OSTI.gov: Livermore, CA, USA, 1987; pp. 1–202. SAND86-7176.
107. Soyama, H. Power law of injection pressure and nozzle diameter on aggressive intensity of a cavitating jet. In Proceedings of the 21st International Conference on Water Jetting, Ottawa, ON, Canada, 19–22 September 2012; pp. 343–354.
108. Hattori, S.; Goto, Y.; Fukuyama, T. Influence of temperature on erosion by a cavitating liquid jet. *Wear* **2006**, *260*, 1217–1223. [CrossRef]
109. Hutli, E.A.F.; Nedeljkovic, M.S.; Radovic, N.A. Mechanics of submerged jet cavitating action: Material properties, exposure time and temperature effects on erosion. *Arch. Appl. Mech.* **2008**, *78*, 329–341. [CrossRef]
110. Dular, M. Hydrodynamic cavitation damage in water at elevated temperatures. *Wear* **2016**, *346*, 78–86. [CrossRef]
111. Kamisaka, H.; Soyama, H. Effect of water depth on aggressive intensity of a cavitating jet. *J. Jet Flow Eng.* **2020**, *35*, 4–11.
112. Soyama, H.; Takakuwa, O.; Naito, A. Effect of nozzle shape for high injection pressure on aggressivity of cavitating jet. In Proceedings of the 21st International Conference on Water Jetting, Ottawa, ON, Canada, 19–22 September 2012; pp. 367–378.
113. Johnson, V.E.; Lindenmuth, W.T.; Conn, A.F.; Frederick, G.S. Feasibility study of tuned-resonator, pulsating cavitating water jet for deep-hole drilling. In *Sandia National Laboratories, Contractor Report SAND81-7126*; OSTI.gov: Livermore, CA, USA, 1981; pp. 1–131.
114. Soyama, H.; Matsuda, K.; Mikami, M. Digging of concrete by a cavitating jet in air comparing with a conventional water jet. In Proceedings of the 23rd International Conference on Water Jetting, Seattle, WA, USA, 16–18 November 2016; pp. 209–214.
115. Saitou, N.; Enomoto, K.; Kurosawa, K.; Morinaka, R.; Hayashi, E.; Ishikwa, T.; Yoshimura, T. Development of water jet peening technique for reactor internal components of nuclear power plant. *J. Jet Flow Eng.* **2003**, *20*, 4–12.
116. Soyama, H. Improvement in fatigue strength of silicon manganese steel SUP7 by using a cavitating jet. *JSME Int. J.* **2000**, *43A*, 173–178. [CrossRef]
117. Soyama, H.; Sasaki, K.; Odhiambo, D.; Saka, M. Cavitation shotless peening for surface modification of alloy tool steel. *JSME Int. J.* **2003**, *46A*, 398–402. [CrossRef]
118. Odhiambo, D.; Soyama, H. Cavitation shotless peening for improvement of fatigue strength of carbonized steel. *Int. J. Fatigue* **2003**, *25*, 1217–1222. [CrossRef]
119. Soyama, H.; Takano, Y.; Ishimoto, M. Peening of forging die by cavitation. In *Technical Review of Forging Technology Institute of Japan*; Forging Technology Insitute of Japan: Tokyo, Japan, 2000; Volume 25, pp. 53–57.
120. Soyama, H.; Macodiyo, D.O. Fatigue strength improvement of gears using cavitation shotless peening. *Tribol. Lett.* **2005**, *18*, 181–184. [CrossRef]
121. Soyama, H.; Sekine, Y. Sustainable surface modification using cavitation impact for enhancing fatigue strength demonstrated by a power circulating-type gear tester. *Int. J. Sustain. Eng.* **2010**, *3*, 25–32. [CrossRef]
122. Soyama, H.; Shimizu, M.; Hattori, Y.; Nagasawa, Y. Improving the fatigue strength of the elements of a steel belt for CVT by cavitation shotless peening. *J. Mater. Sci.* **2008**, *43*, 5028–5030. [CrossRef]
123. Seki, M.; Soyama, H.; Kobayashi, Y.; Gowa, D.; Fujii, M. Rolling contact fatigue life of steel rollers treated by cavitation peening and shot peening. *J. Solid Mech. Mater. Eng.* **2012**, *6*, 478–486. [CrossRef]

124. Soyama, H. Improvement of fatigue strength of duralumin plate with a hole by a cavitating jet. In Proceedings of the 22nd International Conference on Water Jetting, Haarlem, The Netherlands, 3–5 September 2014; pp. 57–65.
125. Soyama, H.; Takeo, F. Comparison between cavitation peening and shot peening for extending the fatigue life of a duralumin plate with a hole. *J. Mater. Process. Technol.* **2016**, *227*, 80–87. [CrossRef]
126. Soyama, H.; Takeo, F. Effect of various peening methods on the fatigue properties of titanium alloy Ti6Al4V manufactured by direct metal laser sintering and electron beam melting. *Materials* **2020**, *13*. paper No. 13-02216. [CrossRef] [PubMed]
127. Soyama, H. Comparison between shot peening, cavitation peening and laser peening by observation of crack initiation and crack growth in stainless steel. *Metals* **2020**, *10*, 63. [CrossRef]
128. Grinspan, A.S.; Gnanamoorthy, R. Effect of nozzle-traveling velocity on oil cavitation jet peening of aluminum alloy, AA6063-T6. *J. Eng. Mater. Technol.Trans. ASME* **2007**, *129*, 609–613. [CrossRef]
129. Lee, H.; Mall, S.; Soyama, H. Fretting fatigue behavior of cavitation shotless peened Ti-6Al-4V. *Tribol. Lett.* **2009**, *36*, 89–94. [CrossRef]
130. Lee, H.; Mall, S. Analysis of fretting fatigue of cavitation shotless peened Ti-6Al-4V. *Tribol. Lett.* **2010**, *38*, 125–133. [CrossRef]
131. Fukuda, S.; Matsui, K.; Ishigami, H.; Ando, K. Cavitation peening to improve the fatigue strength of nitrocarburized steel. *Fatigue Fract. Eng. Mater. Struct.* **2008**, *31*, 857–862. [CrossRef]
132. Takahashi, K.; Osedo, H.; Suzuki, T.; Fukuda, S. Fatigue strength improvement of an aluminum alloy with a crack-like surface defect using shot peening and cavitation peening. *Eng. Fract. Mech.* **2018**, *193*, 151–161. [CrossRef]
133. Takakuwa, O.; Soyama, H. Suppression of hydrogen-assisted fatigue crack growth in austenitic stainless steel by cavitation peening. *Int. J. Hydrog. Energy* **2012**, *37*, 5268–5276. [CrossRef]
134. Kumagai, N.; Takakuwa, O.; Soyama, H. Improvement of delayed fracture resistance on chrome molybdenum steel bolt by cavitation peening. *Trans. JSME* **2016**, *82*, 1–13, paper No. 16-00111.
135. Tsujimura, M.; Shirakashi, K.; Hamada, S.; Hiyama, H.; Tomita, H.; Nadahara, S.; Ota, M. Development of semi-conductor device cleaning by cavitation jet. *Trans. JSME* **2001**, *67B*, 88–94. [CrossRef]
136. Kumano, H.; Sasaki, T.; Soyama, H. Evaluation of the effectiveness of back-side damage gettering in silicon introduced by a cavitating jet. *Appl. Phys. Lett.* **2004**, *85*, 3935–3937. [CrossRef]
137. G32-16 Standard test method for cavitation erosion using vibratory apparatus1. *ASTM Stand.* **2016**, *3*, 1–20.
138. Goetzberger, A.; Shockley, W. Metal precipitates in silicon P-N junctions. *J. Appl. Phys.* **1960**, *31*, 1821–1824. [CrossRef]
139. Tice, W.K.; Tan, T.Y. Nucleation of cusi precipitate colonies in oxygen-rich silicon. *Appl. Phys. Lett.* **1976**, *28*, 564–565. [CrossRef]
140. Kang, J.S.; Schroder, D.K. Gettering in silicon. *J. Appl. Phys.* **1989**, *65*, 2974–2985. [CrossRef]
141. Istratov, A.A.; Hieslmair, H.; Weber, E.R. Iron contamination in silicon technology. *Appl. Phys. A Mater. Sci. Process.* **2000**, *70*, 489–534. [CrossRef]
142. Battaglia, C.; Cuevas, A.; De Wolf, S. High-efficiency crystalline silicon solar cells: Status and perspectives. *Energy Environ. Sci.* **2016**, *9*, 1552–1576. [CrossRef]
143. Mikkelsen, J.C. Gettering of oxygen in Si wafers damaged by ion-implantation and mechanical abrasion. *Appl. Phys. Lett.* **1983**, *42*, 695–697. [CrossRef]
144. Sawada, R. Durability of mechanical damage gettering effect in si wafers. *Jpn. J. Appl. Phys. Part 1 Regul. Pap. Short Notes Rev. Pap.* **1984**, *23*, 959–964. [CrossRef]
145. Soyama, H.; Kumano, H. Oxidation-induced stacking faults introduced by using a cavitating jet for gettering in silicon. *Electrochem. Solid State Lett.* **2000**, *3*, 93–94. [CrossRef]
146. Kumano, H.; Soyama, H. Back side damage gettering of cu using a cavitating jet. *Electrochem. Solid State Lett.* **2004**, *7*, G51–G52. [CrossRef]
147. Kikuchi, M.; Takiguchi, T.; Yamada, J.; Yamamoto, M.; Soyama, H. Evaluation of the efficiency of cleaning dental plaque on titanium using a cavitating jet. *J. Biomech. Sci. Eng.* **2014**, *9*. paper No.14-00297. [CrossRef]
148. Yamada, J.; Takiguchi, T.; Saito, A.; Odanaka, H.; Soyama, H.; Yamamoto, M. Removal of oral biofilm on an implant fixture by a cavitating jet. *Implant Dent.* **2017**, *26*, 904–910. [CrossRef]

149. Yamada, J.; Peng, K.; Kokubun, T.; Takiguchi, T.; Li, G.; Yamamoto, M.; Soyama, H. Optimization of a cavitating jet for removing dental plaque from the surface of the screw of an implant and its practical application. *J. Biomech. Sci. Eng.* **2018**, *13*. paper No.17-00102. [CrossRef]
150. Lindhe, J.; Meyle, J.; European Workshop, P. Peri-implant diseases: Consensus report of the sixth european workshop on periodontology. *J. Clin. Periodontol.* **2008**, *35*, 282–285. [CrossRef]
151. Heitz-Mayfield, L.J.A. Peri-implant diseases: Diagnosis and risk indicators. *J. Clin. Periodontol.* **2008**, *35*, 292–304. [CrossRef] [PubMed]
152. Renvert, S.; Roos-Jansaker, A.M.; Claffey, N. Non-surgical treatment of peri-implant mucositis and peri-implantitis: A literature review. *J. Clin. Periodontol.* **2008**, *35*, 305–315. [CrossRef]
153. Teughels, W.; Van Assche, N.; Sliepen, I.; Quirynen, M. Effect of material characteristics and/or surface topography on biofilm development. *Clin. Oral Implant. Res.* **2006**, *17*, 68–81. [CrossRef] [PubMed]
154. Ley, S.V.; Low, C.M.R. *Ultrasound in Synthesis*; Springer-Verlag: Berlin, Germany, 1980.
155. Sivakumar, M.; Pandit, A.B. Wastewater treatment: A novel energy efficient hydrodynamic cavitational technique. *Ultrason. Sonochem.* **2002**, *9*, 123–131. [CrossRef]
156. Patil, P.N.; Bote, S.D.; Gogate, P.R. Degradation of imidacloprid using combined advanced oxidation processes based on hydrodynamic cavitation. *Ultrason. Sonochem.* **2014**, *21*, 1770–1777. [CrossRef] [PubMed]
157. Bagal, M.V.; Gogate, P.R. Wastewater treatment using hybrid treatment schemes based on cavitation and fenton chemistry: A review. *Ultrason. Sonochem.* **2014**, *21*, 1–14. [CrossRef]
158. Dular, M.; Griessler-Bulc, T.; Gutierrez-Aguirre, I.; Heath, E.; Kosjek, T.; Klemencic, A.K.; Oder, M.; Petkovsek, M.; Racki, N.; Ravnikar, M.; et al. Use of hydrodynamic cavitation in (waste)water treatment. *Ultrason. Sonochem.* **2016**, *29*, 577–588. [CrossRef]
159. Thanekar, P.; Gogate, P. Application of hydrodynamic cavitation reactors for treatment of wastewater containing organic pollutants: Intensification using hybrid approaches. *Fluids* **2018**, *3*, 98. [CrossRef]
160. Gagol, M.; Przyjazny, A.; Boczkaj, G. Wastewater treatment by means of advanced oxidation processes based on cavitation—A review. *Chem. Eng. J.* **2018**, *338*, 599–627. [CrossRef]
161. Bandala, E.R.; Rodriguez-Narvaez, O.M. On the nature of hydrodynamic cavitation process and its application for the removal of water pollutants. *Air Soil Water Res.* **2019**, *12*, 6. [CrossRef]
162. Burzio, E.; Bersani, F.; Caridi, G.C.A.; Vesipa, R.; Ridolfi, L.; Manes, C. Water disinfection by orifice-induced hydrodynamic cavitation. *Ultrason. Sonochem.* **2020**, *60*, 13. [CrossRef]
163. Kalumuck, K.M.; Chahine, G.L. The use of cavitating jets to oxidize organic compounds in water. *J. Fluids Eng.-Trans. Asme* **2000**, *122*, 465–470. [CrossRef]
164. Kato, H.; Oe, Y.; Honoki, M.; Mochiki, T.; Fukazawa, T. Dispersion of spilled oil by a cavitating jet at sea. *J. Mar. Sci. Technol.* **2006**, *11*, 131–138. [CrossRef]
165. Nakashima, K.; Ebi, Y.; Shibasaki-Kitakawa, N.; Soyama, H.; Yonemoto, T. Hydrodynamic cavitation reactor for efficient pretreatment of lignocellulosic biomass. *Ind. Eng. Chem. Res.* **2016**, *55*, 1866–1871. [CrossRef]

Publisher's Note: MDPI stays neutral with regard to jurisdictional claims in published maps and institutional affiliations.

Article

Numerical Insight into the Kelvin-Helmholtz Instability Appearance in Cavitating Flow

Peter Pipp, **Marko Hočevar** and **Matevž Dular** *

Faculty of Mechanical Engineering, University of Ljubljana, Askerceva 6, 1000 Ljubljana, Slovenia; peter.pipp@fs.uni-lj.si (P.P.); marko.hocevar@fs.uni-lj.si (M.H.)
* Correspondence: matevz.dular@fs.uni-lj.si; Tel.: +386-1-477-1453

Abstract: Recently the development of Kelvin-Helmholtz instability in cavitating flow in Venturi microchannels was discovered. Its importance is not negligible, as it destabilizes the shear layer and promotes instabilities and turbulent eddies formation in the vapor region, having low density and momentum. In the present paper, we give a very brief summary of the experimental findings and in the following, we use a computational fluid dynamics (CFD) study to peek deeper into the onset of the Kelvin-Helmholtz instability and its effect on the dynamics of the cavitation cloud shedding. Finally, it is shown that Kelvin-Helmholtz instability is beside the re-entrant jet and the condensation shock wave the third mechanism of cavitation cloud shedding in Venturi microchannels. The shedding process is quasi-periodic.

Keywords: cavitation; Kelvin-Helmholtz instability; microchannel; numerical simulation

Citation: Pipp, P.; Hočevar, M.; Dular, M. Numerical Insight into the Kelvin-Helmholtz Instability Appearance in Cavitating Flow. *Appl. Sci.* **2021**, *11*, 2644. https://doi.org/10.3390/app11062644

Academic Editor: Florent Ravelet

Received: 12 February 2021
Accepted: 12 March 2021
Published: 16 March 2021

1. Introduction

When the pressure drops below the vapor pressure, cavitation, that is, the formation of vapor cavities within a homogeneous liquid medium occurs. The vapor structures are precarious, and they frequently collapse vigorously as they encounter an area of elevated pressure [1]. The mechanism of cavitation cloud shedding is among the most distinctive features of evolved cavitation. Two unique mechanisms are well established to influence the process of cloud shedding [2–7].

- Re-entrant jet: the flow that passes over the attached cavity, deviates towards the surface because of the discrepancies in the pressure in and out of the attached cavity. Just downstream from the attached cavity closure line, a stagnation point forms and divides the flow into a downstream and an upstream portion. The latter enters the attached cavity and causes its separation—creating a detached cavitation cloud that is carried downstream, where it reaches and collapses into a higher-pressure zone. The cycle continues periodically.
- Shock wave: a shock wave that travels through the flow region causes the collapse of the cavitation cloud. It suppresses the attached cavity as it moves upstream. A large vapor cloud is shed when the distortion in the void fraction enters the area of cavity separation at the top of the wedge. The cavity expands again later on, and the cycle continues.

We recently published on the visualization experiment of cavitation within a microchannel [8]. Although the initial objective of the analysis was to create supercavitating conditions within it, we discovered that due to the development of a Kelvin-Helmholtz instability, which induces a semi periodic collapse of the attached cavity, this regime is suppressed. Detailed experimental studies using high-speed photography with visible light and X-rays, showed that this is indeed a third process contributing to the shedding of cloud cavitation in addition to the re-entrant jet and the shock wave mechanisms.

The purpose of this paper is to provide greater insight into the occurrence of Kelvin-Helmholtz instability of cavitation structures in the Venturi microchannel. Above all, we will show how and why the phenomenon occurs, and how it affects the dynamics of cavitation. In doing so, we will focus primarily on showing the results of vapor volume fractions, velocity profiles, and velocity field, thus showing an insight into what is happening not only around and at the boundary, but also within cavitation structures.

We give a very brief summary of the experimental findings in the following sections and then use a computational fluid dynamics (CFD) study to peek deeper into the onset of the Kelvin-Helmholtz instability and its effect on the dynamics of the cavitation cloud shedding.

2. Experimental Observations

The experimental method is seen here only briefly, a more detailed study can be found in [8]. The set-up is illustrated in Figure 1. By sandwiching them between two acrylic glass plates, microchannels are constructed of 450 μm thick stainless-steel sheets that have a laser-cut convergent-divergent narrowing. The Venturi channel's convergent angle measures 18° and the divergent angle is 10°, with a throat height of 675 μm. Both the channel entrance and exit are perpendicular to the cross-section plane, with the channel inlet on the left and the outlet on the right side of the channel.

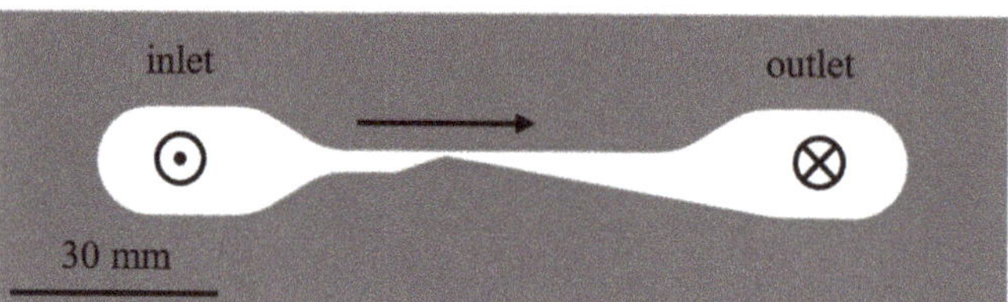

Figure 1. Microchannel geometry with the 675 μm throat gap and the 450 μm channel width.

High-speed cameras (Photron SA-Z and Photron Mini AX200, Photron, Tokio, Japan) have captured cavitation images at a framerate of typically 200,000 fps in both the visible and X-ray light spectrum. Typical flow features that imply the formation of the Kelvin-Helmholtz instability in the channel are shown in Figure 2. The flow is from the left side to the right.

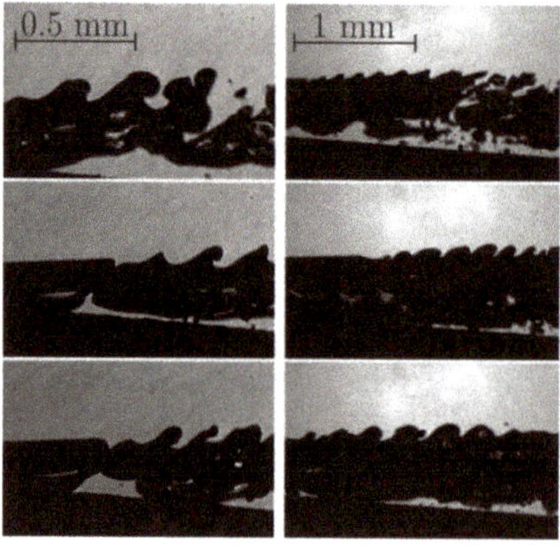

Figure 2. Development of the Kelvin-Helmholtz instability in the microchannel ($\dot{m}$ = 9.15 g/s, Δp = 4.00 bar, σ = 1.24).

Figure 2 illustrates common microchannel representations of the Kelvin-Helmholtz instability. In particular, these images are from an experiment with a mass flow ($\dot{m}$) 9.15 g/s, a pressure difference between inlet and outlet (Δp) 4 bar, and a cavitation number (σ) of 1.24. A closer look at the flow conditions shows that the attached cavity first forms in the shape of a large single cavitation bubble extending downstream from the starting point until the pressure rises well above the saturation pressure—the presence of cavitation is like the supercavitation state. Kelvin-Helmholtz instability forms when the bubble approaches its maximum size. The ripple of the interface increases and the instability encircles the entire cavitation cloud while rolling up and induces its very rapid dissolution in the flow. The pattern then continues periodically.

The numerical simulation that provides further information in the creation of the Kelvin-Helmholtz instability and the cavitation dynamics in the microchannel is presented in the next section.

3. Numerical Procedure

A relatively straightforward simulation was carried out using the commercial Fluent 2020 R2 software package (Ansys Inc, Canonsburg, PA, USA). The solution was obtained with the Unsteady Reynolds-averaged Navier-Stokes (URANS) equations, with Reboud correction modified SST k-ω turbulence model and the Schnerr-Sauer cavitation model. The next section provides some more information.

3.1. Mesh

We regarded the matter two-dimensionally to promote the computation of the case and to expedite the simulation calculation. Thus, as seen in Figure 3, the channel inlet and outlet are modified appropriately. The structured computational mesh was used. The mesh independence was tested on 3 meshes while observing the average pressure difference Δp and the average cavity length l. The discretization error of less than 0.6% was calculated by Richardson extrapolation [9] for the medium mesh with around 160,000 cells at a flow rate of $\dot{m}$ = 9.03 g/s (Table 1).

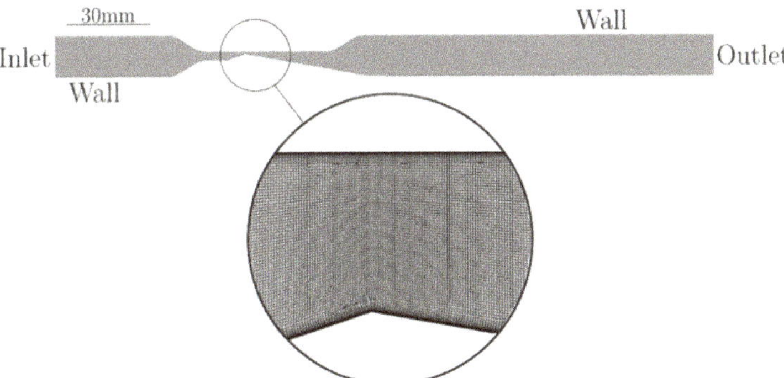

Figure 3. The geometry of the microchannel computational domain and the mesh detail in the wedge apex area.

Table 1. Mesh independence study.

Mesh Size	Δp (bar)	l (mm)
~80,000	3.56	25.0
~160,000	3.71	25.6
~320,000	3.73	25.7

Along with the channel height, the final mesh had a constant number of 67 cells. Therefore, 10 μm is the height of an individual cell at the wedge apex. In relation to the geometry, the cells grow in the length to height ratio towards the inlet and the outlet of the domain. We have considered the boundary layer, where the overall thickness of the boundary layer of 50 μm was defined, with 10 layers and a growth rate of 1.2, which is a sufficiently small mesh to maintain the y+ values below 5.

3.2. Reynolds-Averaged Navier-Stokes (RANS) Equations

The Navier-Stokes equations are a set of partial differential equations, describing the conservation of momentum and mass for a viscous fluid flow motion. A direct numerical simulation, in which the Navier-Stokes equations are solved without a turbulence model, requires a huge amount of computing power because the whole range of spatial and temporal scales of the turbulence must be resolved. As this is not practical, averaging of variables is used. The idea behind averaging is the Reynolds decomposition, an instantaneous quantity decomposition into its time-averaged Φ and fluctuating quantities φ′(t) with zero mean value. Reynolds-averaged Navier-Stokes equations will be used in the form, that is generally accepted in the commercially available computational fluid dynamics (CFD) solvers. Here, the overline indicates a time-averaged variable, while the tilde indicates a density-weighted averaged or Favre-averaged variable [10]. The Favre-averaging method was selected because in the present case turbulent fluctuations lead to sizeable density fluctuations. The continuity equation is as follows, where ρ is the fluid density, t time, and $\boldsymbol{U}$ velocity vector, defined as $\boldsymbol{U} = [U, V, W]^T$:

$$\frac{\partial \overline{\rho}}{\partial t} + \nabla \cdot \left(\overline{\rho} \tilde{\mathbf{U}} \right) = 0 \tag{1}$$

The momentum equations in the x and y directions are as follows, where P is pressure, μ dynamic viscosity, u', v' and w' velocity fluctuations of velocity vector $\boldsymbol{U}$, while S_{Mx} and S_{My} represent momentum source terms in x and y directions, respectively

$$\begin{aligned} \frac{\partial(\overline{\rho}\overline{U})}{\partial t} + \nabla \cdot (\overline{\rho}\overline{U}\boldsymbol{U}) &= -\frac{\partial \overline{P}}{\partial x} + \nabla \cdot (\mu \nabla \overline{U}) + \left[-\frac{\partial\left(\overline{\rho u'^2}\right)}{\partial x} - \frac{\partial(\overline{\rho u'v'})}{\partial y} - \frac{\partial(\overline{\rho u'w'})}{\partial z} \right] + S_{Mx} \\ \frac{\partial(\overline{\rho}\overline{V})}{\partial t} + \nabla \cdot (\overline{\rho}\overline{V}\boldsymbol{U}) &= -\frac{\partial \overline{P}}{\partial y} + \nabla \cdot (\mu \nabla \overline{V}) + \left[-\frac{\partial(\overline{\rho u'v'})}{\partial x} - \frac{\partial\left(\overline{\rho v'^2}\right)}{\partial y} - \frac{\partial(\overline{\rho v'w'})}{\partial z} \right] + S_{Mx} \end{aligned} \tag{2}$$

For the accurate representation of cavitating flow details, some authors perform compressible flow simulations [11–13]. In the compressible case, the energy conservation equation is added to the above set of continuity and momentum equations. Also, equations of state of vapor and liquid are introduced. On the other hand, some authors have shown, that sufficient flow representation accuracy was obtained in the incompressible flow case, much decreasing the computational load requirements (for example [2,14]).

3.3. Turbulence Modeling

Increased available computational power has made possible advances in computational fluid dynamics, among them Large-Eddy Simulation (LES) and Detached-Eddy Simulation (DES). These relatively novel approaches are in research used to accurately resolve all scales of turbulent fluctuations, while for engineering purposes there is usually no need for such complexity. In above mentioned Reynolds-averaging models, turbulent fluctuations are not resolved, and the turbulence model is used instead. The k-ω and k-ε turbulence models are the most common models used to simulate mean flow characteristics for turbulent flow conditions. Both are two-equation models that provide a general description of turbulence utilizing two transport equations. The Menter's Shear Stress Transport (SST) k-ω turbulent model [15], combines both k-ω and k-ε turbulence models such that it merges the robust, accurate, and less y+ sensitive performance of the k-ω model in the near-wall region with the solid freestream operation of the k-ε model in the far-field. Nowadays the SST k-ω turbulent model is the preferred two-equation turbulent model and the preferred model for a wide range of applications.

For the modeling of nonstationary cavitation flows, among them, periodic cavitation cloud shedding, two-equation turbulent models in the original form are not the preferred option. Two equation models prevent the reentrant jet from appearing at the rear of the cavitation cloud, all this being caused by too high prediction of turbulent viscosity. For a better non-stationary turbulence modeling performance, Reboud et al. [16] proposed a modification of the turbulent model. In the modification, the turbulent viscosity of the mixture is reduced in regions with high void fractions and the equation of turbulent viscosity becomes as follows, with k representing turbulence kinetic energy and ω specific turbulence dissipation rate:

$$\mu_t = f(\rho)\frac{k}{\omega} \tag{3}$$

The density function is set by Equation (4), here indices l, v, m represent liquid, vapor, and mixture.

$$f(\rho) = \rho_v + \frac{(\rho_m - \rho_v)^n}{(\rho_l - \rho_v)^{n-1}} \; n \gg 1 \tag{4}$$

Various values were proposed for the exponent n. Coutier-Delgosha et al. [17] proposed value n = 10, while the described turbulent model extension was used and validated for various applications, among them in Venturi channel and a hydrofoil [2,3,18–20].

3.4. Two-Phase Flow Modeling

We more commonly use the concept of the homogeneous flow of the mixture for cavitation simulation, where the two-phase flow is assumed to be a single-phase flow of the liquid-vapor mixture. This helps one to solve only one motion equation since we consider the problem as a single-phase, but with the mixture's properties. Accordingly, the properties of the liquid-vapor mixture are described by the proportion of the vapor phase, using the model suggested by Bankoff [21]. The mixture's density is written:

$$\rho_m = \alpha\rho_v + (1-\alpha)\rho_l \tag{5}$$

where α denotes vapor volume fraction, and mixture's dynamic viscosity is written as:

$$\mu_m = \alpha\mu_v + (1-\alpha)\mu_l \tag{6}$$

The equations of mass and momentum conservation are resolved in the model of the homogeneous flow of the mixture by the properties of the mixture, and the equation of phase fraction conservation must be solved:

$$\frac{\partial}{\partial t}(\rho_v\alpha) + \nabla\cdot(\rho_v\alpha\mathbf{u}_v) = R_e - R_c \tag{7}$$

where R_e and R_c represent mass transfer source terms for evaporation and condensation, which account for the mass transfer between the phases of liquid and vapor in cavitation and are thus related to the growth and collapse of the vapor bubbles. According to the cavitation model used, their formulation varies.

3.5. Cavitation Model

We know multiple models of cavitation in which we can model liquid and vapor evaporation and condensation. Bubble dynamics models, introduced by Kubota et al. [22], where he used a linear portion of the Rayleigh-Plesset equation to explain the evolution of bubble radius as a function of surrounding pressure have been the most used cavitation models in CFD in recent years. The proportion of the vapor phase and hence the density of the mixture is calculated by bubble radius and bubble number density. Other cavitation models were derived relying on the pressure and bubble radius dependencies [23], based on the Rayleigh-Plesset equation. However, with all cavitation models, it is conditioned to specify parameters, such as the bubble number density or the initial size of the bubble, which are very hard to define. The Schnerr-Sauer [24] model is the cavitation model based on bubble dynamics that is most widely used in commercial CFD software packages, where the mass transfer source terms R_e and R_c (see Equation (7) are defined as:

when local pressure is equal to or smaller than vapor pressure—$p \leq p_v$:

$$R_e = F_{evap} \frac{\rho_v \rho_l}{\rho} \alpha(1-\alpha) \frac{3}{R_b} \sqrt{\frac{2}{3} \frac{(p_v - p)}{\rho_l}} \quad (8)$$

when local pressure is larger than vapor pressure—$p > p_v$:

$$R_c = F_{cond} \frac{\rho_v \rho_l}{\rho} \alpha(1-\alpha) \frac{3}{R_b} \sqrt{\frac{2}{3} \frac{(p - p_v)}{\rho_l}} \quad (9)$$

where the default values of empirical calibration coefficients of evaporation F_{evap} and condensation F_{cond} are 1 and 0.2, respectively. To link the fraction of the vapor volume to the number of bubbles per liquid volume n_b, the Schnerr-Sauer cavitation model uses:

$$\alpha = \frac{n_b \frac{4}{3} \pi R_b^3}{1 + n_b \frac{4}{3} \pi R_b^3} \quad (10)$$

The parameters that must be determined in this model are bubble radius R_b and bubble number density n. The compressibility of water and water vapor was not considered in the modeling.

3.6. Boundary Conditions

Modeling cavitating flow in the microchannel was performed at several mass flow rates. The boundary condition was set at the inlet of the Venturi channel computational domain as average velocity. The inlet turbulence intensity was set to zero. The vapor volume fraction at the inlet was set to zero also.

At the outlet of the Venturi channel computational domain the second boundary condition was set prescribing the absolute pressure 1 bar and backflow turbulence intensity 5% for all mass flow rates. The vapor volume fraction at the outlet was set to zero.

On all walls, the no-slip stationary wall boundary condition was used with a standard wall roughness model. The temperature was in all cases set to 20 °C.

3.7. Physics and Solver Settings

It is important to set proper simulation settings, as are the initial conditions from which to start the simulation, so we first performed simulations as single-phase simulations for each mass flow under stationary conditions. The outcomes were then considered as the initial conditions for further transient simulations. Numerical calculations were carried out using time-dependent Reynolds-averaged Navier-Stokes equations. Homogeneous water and water vapor mixture was considered and a Schnerr-Sauer cavitation model [24] with an evaporation pressure of 2340 Pa was used. A modified SST k-ω model, using the turbulent viscosity correction suggested by Reboud et al. [16] and Coutier-Delgosha et al. [17], mentioned above, was used for the turbulent model. For pressure and velocity coupling, the PISO algorithm [25] was used. We used the second-order upwind scheme [26] for spatial numerical discretization of solving hyperbolic partial differential equations for all but the pressure and volume fraction, providing more precise results with marginally higher computer resource usage compared to the first-order upwind scheme. The PRESTO interpolation scheme was used to discretize the pressure [27], and the first-order upwind scheme [26] was used for the discretization of the volume fraction. Time integration was modeled for transient solutions using the implicit transient formulation of bounded second-order.

The convergence criteria were established by evaluating the progression of various flow parameters (absolute pressure at the inlet and outlet, and velocities at the microchannel inlet, outlet, and throat). Following the sum of the imbalance of transport equations between iterations for all cells in the computational domain, the observed flow parameters were converged when falling below 10^{-5} of the iterative numerical solution of the individual equations at each time step of the simulation. The iteration error was estimated to be less than 0.02%. By assessing its effect against the average pressure difference and the cavity length, the size of the time step was obtained. There was little difference in these parameters if the time step was shorter than 5 µs, but a shorter one—1 µs was ultimately chosen for observation of the Kelvin-Helmholtz instability. We conducted 50 ms of computational modeling for each case, where the last 30 ms was applicable for further analysis.

4. Results

First, we assess the capability of the numerical simulation to capture the general cavitation dynamics which was observed inside the microchannel—more specifically the onset of the Kelvin-Helmholtz instability (Figure 4). The flow is from the left to the right, the time difference between the images is 100 µs.

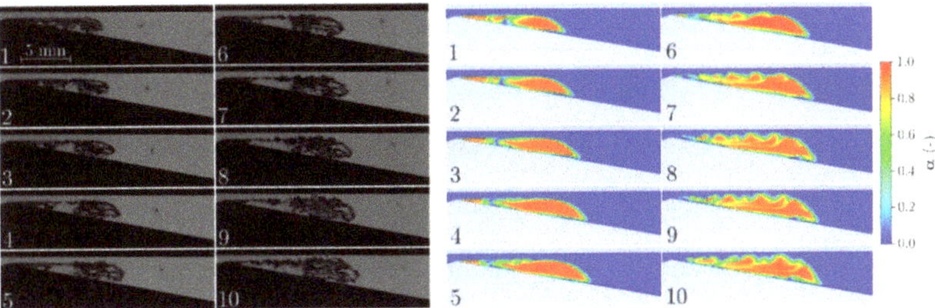

Figure 4. Separation of cavitation clouds and Kelvin-Helmholtz instability formation (experiment: $\dot{m}$ = 9.15 g/s, Δp = 4.00 bar, σ = 1.24; simulation: $\dot{m}$ = 9.03 g/s, Δp = 3.71 bar, σ = 1.26). The time difference between the images is Δt = 100 µs.

The experiment and simulation both reveal the same story, but in the case of simulation, the Kelvin-Helmholtz instability is more pronounced and can be understood better and described more in detail.

In Figure 5, the development of the Kelvin-Helmholtz instability is shown with velocity profiles drawn over at cross-sections at horizontal intervals of 3 mm in the downstream direction, starting at the wedge apex.

Fluid flow velocity at the throat of the Venturi microchannel is 30 m/s, where liquid flows over the cavity and descents afterward towards the bottom wall of the channel, where the flow divides into the upstream and downstream flow (1). Vapor fraction α profiles for both phases are approximately uniform in both the fluid and vapor layers. They are however discontinuous at the straight interface between the two (step 1). Shear stress acts on the fluids due to discontinuity of the velocity gradient (velocity gradients are shown by the black line in Figure 5), while vapor gradient locally promotes the generation of instabilities and vorticity. The backflow due to the presence of the adverse pressure gradient at 6 mm downstream from the throat in the vapor phase increases from a peak value of 5 m/s to 13 m/s (step 2), which further increases shear stress. Increased shear stress between the liquid and cavity interface results in the formation of ripples on the interface, while the reentrant jet progresses further upstream. The attached cavity expands and reaches its full size (step 2) and shortly thereafter (100 µs) in step (3) one can see the first beginnings of ripples on the upper edge area of the vapor phase (approx. 6 mm from the Venturi throat). This promoting the formation of instabilities (step 3) in addition to the already present vertical shear stress between the liquid flow and the cavity interface enables the formation of Kelvin Helmholtz instability.

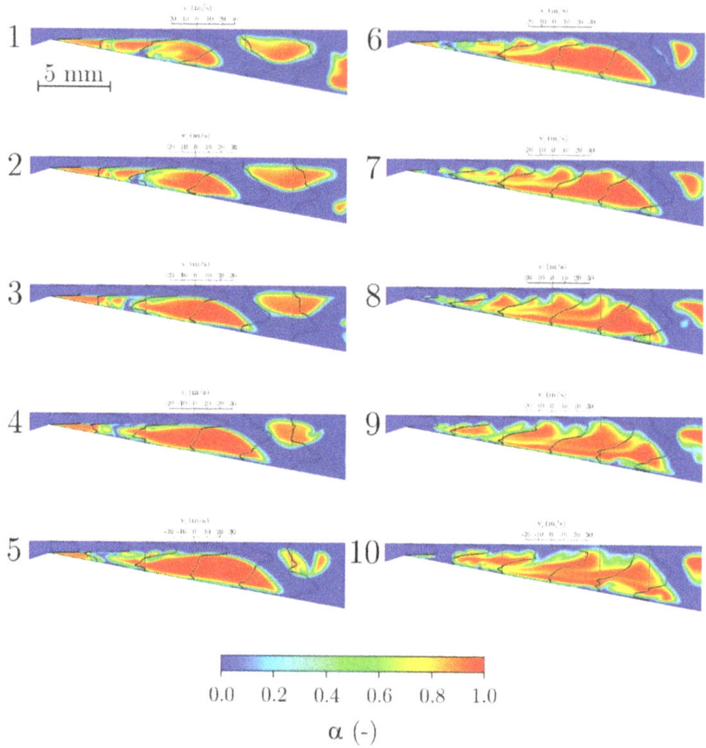

Figure 5. Development of Kelvin-Helmholtz instability with velocity profiles at individual cross-sections. The time difference between the images is Δt = 100 μs.

In step (4), 300 μs after the start of the instability formation process, we already observe the formation of unstable local turbulent structures approx. 6 mm from the Venturi throat, which are now confined to the shear layer. Due to their limited size, velocity gradients in step (4) are still unchanged in comparison to step (1) and the rolling motion of turbulent structures is not yet recognized. With time, the surface becomes more and more deformed from the original straight boundary interface, and turbulent structures grow in size and number (step 5). It seems that the vapor region is affected by the formation of turbulent structures, here velocity, density, and momentum are low. Velocity profiles in the vapor region change with instabilities continuously protruding deeper in the vapor region, finally leading to the situation in step (5).

Instabilities in the vapor phase continue to grow from step (5), and the region of high pressure between the cavities moves in the opposite direction to the main flow towards the Venturi throat, from approx. 4 mm in step (5) to approx. 2 mm in step (6). Advance in the direction opposite is related to the presence of the adverse pressure gradient (step 6), with instabilities and turbulent eddies in later steps reaching the Venturi throat section and causing full cavity separation and partial destruction of the vapor region in the vicinity of the throat. The vapor phase in the vicinity of the throat does not reappear until step (10) 900 μs after the start of the process. In steps (7–9) we notice the change of the velocity profiles, caused by the separation of the cavity from the Venturi throat and hence smoothing of the velocity profiles in the horizontal direction. The region of constant velocity near the upper Venturi section wall is reduced and the velocity profiles in horizontal direction become skewed. Regions of backflow are reduced in size and intensity. In steps (7–9) generation of Kelvin Helmholtz instabilities increase further downstream of the detached cavitation cavity. Finally, the entire detached cavitation cloud becomes engulfed in the border instabilities. In addition to the change of velocity gradient, the vapor fraction α in the vertical direction becomes non-homogeneous and areas of slightly higher vapor fraction α can be traced to the region where instabilities were initially formed. The detached cavitation cloud together with all instabilities continues to flow downstream and finally dissipates (not shown in this sequence).

In step (10), 900 µs after the start of the instabilities formation, cavitation cavity is again formed at the Venturi throat section and the entire process is about to continue in a quasi-periodic way. More insight into the instability formation process may be gained from Figure 6, which shows an enlarged segment of the Venturi section relative to Figure 5.

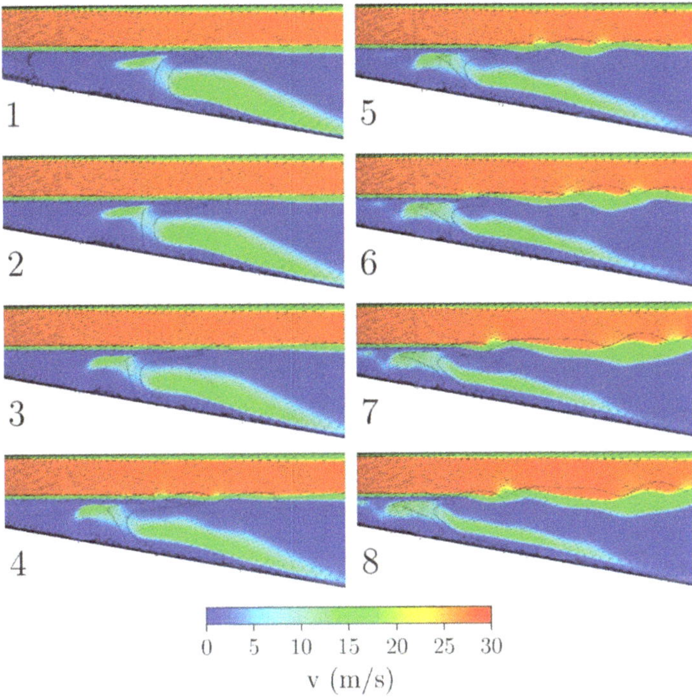

Figure 6. A close-up view of the initial development of Kelvin-Helmholtz instability. The time difference between the images is $\Delta t = 20$ µs. The black line shows a 10% vapor fraction region limit.

Velocity vectors reveal the presence of the turbulent eddies within the vapor phase and their interaction with the main flow above the discontinuity of the horizontal velocity. Figure 6 also shows contours of 10% vapor region location.

Step (1) shows the initial situation with a turbulent eddy on the right side of the observation region. Just next to the lower limit of 10% vapor region is the turbulent eddy impingement point on the shear layer. On the left of the 10% vapor fraction region, we observe the backflow, while on the right the turbulent eddy rotation causes the flow velocity to be in the direction of the main flow. Impingement point marks the position of the large variation of the shear stress gradient, on the left of the impingement point, the stress is higher than on the right. The horizontal 10% vapor fraction region limit of the velocity discontinuity is just above the impingement point. The transition from step (1) to step (2) is marked with an additional depression of the horizontal 10% vapor fraction region limit. The location of the impingement point and the horizontal 10% vapor fraction region has moved to the right in step (2) relative to step (1), while the lower 10% vapor fraction region has widened from the original location in step (1) to the newly established impingement point of step (2).

As the instability process continues in step (3) we observe gradual folding of the lower 10% vapor fraction limit, even before we notice significant variations in the velocity within the shear layer. We, therefore, speculate that the vapor fraction properties influence the instability process through the variation of the flow density and the transfer of the momentum. In step (3) we notice velocity variations within the main flow, corresponding well to the vapor fraction variations in intensity and location. The process, first observed in step (3) continues with step (4) showing increased spatial variations of the vapor fraction, followed by velocity variations in the main flow. As a consequence, turbulent eddy, first responsible for impinging the shear layer, is decayed into two turbulent eddies with an additional one formed near the location of the original impingement point. The location of the lower 10% vapor fraction region limit has moved in the direction of the Venturi section throat.

Steps (5) to (8) show markedly increasing folding of the shear layer, coupled with the additional breaking of the turbulent eddy into several smaller ones, including interactions among neighboring turbulent eddies. Gradually location of the lower 10% vapor fraction region moves further left towards the throat finally causing the detachment of the cavitation cloud as described above for Figure 5. Folding of the shear layer is coupled with its gradual thickening. Location, where the gradient of shear layer folding is high, is accompanied by the regions of main flow deceleration and hence variations of the velocity and pressure in the main flow.

5. Conclusions

Numerical simulation has provided additional insights into the formation of Kelvin-Helmholtz flow instabilities in the Venturi section. The process is being able to cause the shedding of cavitation clouds similar to the re-entrant jet or shockwave cases. Results conform well to the available experimental data.

The formation of Kelvin-Helmholtz instabilities is determined by the large vertical variations in the shear stress, vapor fraction, and the different physical properties of liquid and vapor regions. The vapor with its much lower density and momentum is more impacted by the instabilities creation as the liquid region. This promotes perturbations and the development of the turbulent eddies in the vapor region. Turbulent eddy impingement point on the shear layer point marks the position of the large variation of the shear stress gradient. We observed the folding of the lines of constant vapor fraction within the shear layer even before any velocity changes were noticed. Thus, the vapor fraction properties may influence the Kelvin-Helmholtz instability formation process through the variation of the density and the transfer of the momentum. As instabilities grow further, adverse pressure gradient and velocity lead to the detachment of the cavitation cloud together with all instabilities.

In both layers, Kelvin-Helmholtz instabilities for the case of cavitation in the Venturi section express less pronounced rolling motion and are less periodic and symmetric in comparison with their more well-known occurrences. This perhaps was the reason that Kelvin-Helmholtz instability was not recognized before as a possible source of the cavitating flow instabilities and cavitation cloud shedding.

More experimental and numerical studies will be required in the future to better understand the formation of Kelvin-Helmholtz flow instabilities in cavitation flow and their role in the cavitation shedding process.

Author Contributions: Conceptualization, M.D. and P.P.; methodology, P.P., M.H. and M.D.; validation, P.P., M.H. and M.D.; Investigation, P.P., M.H. and M.D.; resources, M.H. and M.D.; writing—original draft preparation, P.P., M.H. and M.D.; writing—review and editing, P.P., M.H. and M.D.; visualization, P.P. and M.D.; supervision, M.H. and M.D.; funding acquisition, M.H. and M.D. All authors have read and agreed to the published version of the manuscript.

Funding: European Research Council (ERC) under the European Union's Framework Program for research and innovation, Horizon 2020 (grant agreement n 771567—CABUM), the Slovenian Research Agency (Core funding No. P2-0401).

Institutional Review Board Statement: Not applicable.

Informed Consent Statement: Not applicable.

Acknowledgments: The authors acknowledge the financial support from the European Research Council (ERC) under the European Union's Framework Program for research and innovation, Horizon 2020 (grant agreement n 771567—CABUM), the Slovenian Research Agency (Core funding No. P2-0401) and the PhD grant for young researchers (P.P.).

Conflicts of Interest: The authors declare no conflict of interest.

References

1. Abbas-Shiroodi, Z.; Sadeghi, M.T.; Baradaran, S. Design and optimization of a cavitating device for Congo red decolorization: Experimental investigation and CFD simulation. *Ultrason. Sonochem.* **2021**, *71*. [CrossRef] [PubMed]
2. Dular, M.; Bachert, R.; Stoffel, B.; Širok, B. Experimental evaluation of numerical simulation of cavitating flow around hydrofoil. *Eur. J. Mech. B/Fluids* **2005**, *24*, 522–538. [CrossRef]
3. Dular, M.; Bachert, R.; Schaad, C.; Stoffel, B. Investigation of a re-entrant jet reflection at an inclined cavity closure line. *Eur. J. Mech. B/Fluids* **2007**, *26*, 688–705. [CrossRef]
4. Ganesh, H.; Makiharju, S.A.; Ceccio, S.L. Bubbly shock propagation as a mechanism for sheet-to-cloud transition of partial cavities. *J. Fluid Mech.* **2016**, *802*, 37–78. [CrossRef]
5. Ganesh, H.; Mäkiharju, S.A.; Ceccio, S.L. Bubbly shock propagation as a mechanism of shedding in separated cavitating flows. *J. Hydrodyn.* **2017**, *29*, 907–916. [CrossRef]
6. Laberteaux, K.R.; Ceccio, S.L. Partial cavity flows. Part 1. Cavities forming on models without spanwise variation. *J. Fluid Mech.* **2001**, *431*, 1–41. [CrossRef]
7. Laberteaux, K.R.; Ceccio, S.L. Partial cavity flows. Part 2. Cavities forming on test objects with spanwise variation. *J. Fluid Mech.* **2001**, *431*, 43–63. [CrossRef]
8. Podbevšek, D.; Petkovšek, M.; Ohl, C.D.; Dular, M. Kelvin-Helmholtz Instability Governs the Cavitation Cloud Shedding in Microchannels. *Int. J. Multiph. Flow* **2021**. Submitted.
9. Ferziger, J.H.; Perić, M. *Computational Methods for Fluid Dynamics*; Springer: Berlin/Heidelberg, Germany, 2002.
10. Versteeg, H.K.; Malalasekera, W. *An Introduction to Computational Fluid Dynamics—The Finite Volume Method*; Pearson: London, UK, 2007.
11. Schreiner, F.; Paepenmöller, S.; Skoda, R. 3D flow simulations and pressure measurements for the evaluation of cavitation dynamics and flow aggressiveness in ultrasonic erosion devices with varying gap widths. *Ultrason. Sonochem.* **2020**, *67*. [CrossRef] [PubMed]
12. Mottyll, S.; Skoda, R. Numerical 3D flow simulation of ultrasonic horns with attached cavitation structures and assessment of flow aggressiveness and cavitation erosion sensitive wall zones. *Ultrason. Sonochem.* **2016**, *31*, 570–589. [CrossRef]
13. Mihatsch, M.S.; Schmidt, S.J.; Adams, N.A. Cavitation erosion prediction based on analysis of flow dynamics and impact load spectra. *Phys. Fluids* **2015**, *27*. [CrossRef]
14. Arabnejad, M.H.; Svennberg, U.; Bensow, R.E. Numerical assessment of cavitation erosion risk using incompressible simulation of cavitating flows. *Wear* **2021**, *464–465*, 203529. [CrossRef]
15. Menter, F.R. Two-equation eddy-viscosity turbulence models for engineering applications. *AIAA J.* **1994**, *32*, 1598–1605. [CrossRef]
16. Reboud, J.-L.; Stutz, B.; Coutier, O. Two-phase flow structure of cavitation: Experiment and modelling of unsteady effects. *Phys. Fluids* **1998**, *2017*, 203–208.
17. Coutier-Delgosha, O.; Fortes-Patella, R.; Reboud, J.L. Evaluation of the turbulence model influence on the numerical simulations of unsteady cavitation. *J. Fluids Eng.* **2003**, *125*, 38–45. [CrossRef]
18. Coutier-Delgosha, O.; Reboud, J.L.; Delannoy, Y. Numerical simulation of the unsteady behaviour of cavitating flows. *Int. J. Numer. Methods Fluids* **2003**, *42*, 527–548. [CrossRef]
19. Goncalves, E. Numerical study of unsteady turbulent cavitating flows. *Eur. J. Mech. B/Fluids* **2011**, *30*, 26–40. [CrossRef]
20. Ji, B.; Luo, X.W.; Arndt, R.E.A.; Wu, Y. Numerical simulation of three dimensional cavitation shedding dynamics with special emphasis on cavitation-vortex interaction. *Ocean Eng.* **2014**, *87*, 64–77. [CrossRef]
21. Bankoff, S.G. A variable density single-fluid model for two-phase flow with particular reference to steam-water flow. *J. Heat Transfer* **1960**, *82*, 265–272. [CrossRef]
22. Kubota, A.; Kato, H.; Yamaguchi, H. A new modelling of cavitating flows: A numerical study of unsteady cavitation on a hydrofoil section. *J. Fluid Mech.* **1992**, *240*, 59–96. [CrossRef]
23. Frobenius, M.; Schilling, R. Three-Dimensional Unsteady Cavitating Effects on a Single Hydrofoil and in a Radial Pump Measurement and Numerical Simulation. In Proceedings of the CAV2003, Boulder, CO, USA, 8–12 July 2003; pp. 1–7.
24. Schnerr, G.H.; Sauer, J. Physical and numerical modeling of unsteady cavitation dynamics. In Proceedings of the Internationl Conference on Multiphase Flow, New Orleans, LA, USA, 27 May–1 June 2001; pp. 1–12.
25. Issa, R.I. Solution of the implicitly discretised fluid flow equations by operator-splitting. *J. Comput. Phys.* **1986**, *62*, 40–65. [CrossRef]

26. Barth, T.J.; Jespersen, D. The design and application of upwind schemes on unstructured meshes. In Proceedings of the 27th Aerospace Sciences Meeting, Reno, NV, USA, 9–12 January 1989. [CrossRef]
27. Patankar, S.V. *Numerical Heat Transfer and Fluid Flow*; Hemisphere Publishing Corp: London, UK, 1980.

Article

Development of Attached Cavitation at Very Low Reynolds Numbers from Partial to Super-Cavitation

Florent Ravelet *, Amélie Danlos, Farid Bakir, Kilian Croci, Sofiane Khelladi and Christophe Sarraf

Arts et Metiers Institute of Technology, CNAM, LIFSE, HESAM University, 75013 Paris, France; amelie.danlos@ensam.eu (A.D.); farid.bakir@ensam.eu (F.B.); kilian.croci@ensam.eu (K.C.); sofiane.khelladi@ensam.eu (S.K.); christophe.sarraf@ensam.eu (C.S.)

* Correspondence: florent.ravelet@ensam.eu

Received: 1 October 2020; Accepted: 16 October 2020; Published: 20 October 2020

Abstract: The present study focuses on the inception, the growth, and the potential unsteady dynamics of attached vapor cavities into laminar separation bubbles. A viscous silicon oil has been used in a Venturi geometry to explore the flow for Reynolds numbers ranging from $Re = 800$ to $Re = 2000$. Special care has been taken to extract the maximum amount of dissolved air. At the lowest Reynolds numbers the cavities are steady and grow regularly with decreasing ambient pressure. A transition takes place between $Re = 1200$ and $Re = 1400$ for which different dynamical regimes are identified: a steady regime for tiny cavities, a periodical regime of attached cavity shrinking characterized by a very small Strouhal number for cavities of intermediate sizes, the bursting of aperiodical cavitational vortices which further lower the pressure, and finally steady super-cavitating sheets observed at the lowest of pressures. The growth of the cavity with the decrease of the cavitation number also becomes steeper. This scenario is then well established and similar for Reynolds numbers between $Re = 1400$ and $Re = 2000$.

Keywords: partial cavitation; super-cavitation; laminar cavitation; cavitation instabilities

---

1. Introduction

The presence of small separation bubbles close to the leading edge of a profile or to the summit of a wedge provides favorable conditions for the attachment of vaporous cavities when lowering the absolute pressure. The fundamentals of sheet cavitation inception can be found for instance in the classical books of C. E. Brennen [1] or J.-P. Franc [2]. The inception of such "sheet" cavitation has been for instance studied in Refs. [3–7] on smooth axi-symmetric bodies, on propellers blades or on foils. Moreover, a recent review of the different sheet cavitation inception mechanisms can be found in Ref. [8].

Once the inception of cavitation has occurred, for lower pressures the attached sheet cavities usually grow and become unstable, leading to periodical cloud shedding [9–11]. Partial and cloud cavitations are associated with a vapor region that extends over a part of the cavitating body [2]. At very low pressures, one can observe super-cavitation, consisting of a large and stable vapor cavity that extends beyond the body and closes in the liquid. In that case the pressure fluctuations are usually small [12,13]. For turbomachinery applications, the unsteady dynamics of attached sheets may be responsible for an increase of noise and vibrations and is of crucial importance for system instabilities [14,15]. The instabilities of attached sheet cavitation are attributed to two main phenomena: the formation of a re-entrant jet which is governed by inertia [16], and a bubbly shock propagation mechanism [17]. These two mechanisms have been recently subject to extensive studies, at least for very large Reynolds numbers, in water [18–22]. The cavitation in that case arises on a turbulent flow background.

The current studies that are conducted in the LIFSE (Laboratory of Fluid Engineering and Energetic Systems) deal with cavitating flows at very low Reynolds numbers, i.e., in laminar or transitional regimes. To the best of our knowledge, only very few studies have been carried out on hydrodynamic cavitation [23,24] or flow-induced degassing [25,26] in laminar flows. The main goal of the present study is to identify and compare potential unsteady regimes in viscosity-dominated cavitating flows to the aforementioned regimes observed in turbulent flows. In the present study, viscous silicone oil is used in a well-documented Venturi-type geometry [10]. In a first study conducted in our laboratory (Croci et al. (2018, 2019) [27,28]), the main features of the multiphase structures that arise have been observed from a qualitative point of view, using silicone oil. Silicone oil can dissolve much more air than water under the same thermodynamical conditions [29,30]. In the work of Croci et al. (2018, 2019) [27,28] the oil was saturated with air at atmospheric pressure. Single-phase numerical simulations have been performed additionally to illustrate the flow topology and to estimate the threshold for cavitation inception. As a result, different types of cavities are observed: tadpoles that attach on the lateral wall [27], sheets that attach on the Venturi slope and top-tadpoles that attach on the opposite side of the vein. The numerical simulations, pressure measurements and high-speed visualizations suggest to attribute the tadpoles to degassing, the main cavity sheet to cavitation, and the top-tadpoles to a secondary flow separation that captures air [28].

In the present article, experiments were performed with degassed silicone oil, to remove air bubbles and to get rid of degassing phenomena [25,26,31]. As a result, only central attached sheet cavities are present in the flow. Quantitative measurements based on high-speed visualizations are performed in order to characterize the evolution of the shapes, lengths and temporal features of the attached sheets with a decrease of the ambient pressure at constant Reynolds numbers Re, both in steady laminar flow (at $Re \leqslant 1200$) and in transitional flows (up to $Re \simeq 2000$).

Section 2 is dedicated to the presentation of the experimental setup. The test-bench and the protocol are described in Section 2.1; the main parameters and the post-processing of the high-speed movies are discussed in Section 2.2; and a few numerical results giving insight into the single-phase flow topology and about its transition are recalled in Section 2.3.

The main results are presented in Section 3. The differences between experiments with air-saturated and degassed oil are illustrated in Section 3.1. The different regimes are presented in Section 3.2 and are quantitatively characterized in Section 3.3. Concluding remarks are then given in Section 4.

2. Experimental Study Overview

2.1. Experimental Test-Bench and Protocol

The experiments are conducted in a test-bench of the *LIFSE* facilities especially designed to study cavitation in silicone oils (see Figure 1). A Pollard MPLN 142 volumetric pump (1) impels the silicone oil in the loop. A tank of 100 liters (2) is equipped with a temperature sensor ThermoEst PT100 (3). The tank is connected to compressed air supply or to a vacuum pump (4). The oil then flows in a 40 mm inner diameter pipe and through an ultrasonic flowmeter KHRONE Optisonic 3400 (5) placed 1.5 m upstream of the test section (7). The inlet and outlet pressures are monitored with two JUMO dTrans p30 pressure sensors (6).

The test section consists of a Venturi geometry with convergent/divergent angles of respectively 18° and 8° [10]. It presents an inlet section $S_{in} = 20 \times 10$ mm^2 and a square section at the throat $S_{throat} = 10 \times 10$ mm^2. Both side and top visualizations are possible. Two round-to-rectangle contraction nozzles with an area ratio of 6.3 are placed on each sides of the test section to connect it to the pipe system.

All experiments are realized following a protocol to eliminate a maximum of dissolved air. The pressure in the setup is first decreased to an absolute pressure of $\simeq$70 mbar, then the oil is kept circulating for this pressure at slow velocity during two hours, while removing periodically the air that accumulates in the vertical tube situated above the outlet of the test section and visible on the top part of the sketch on the left in Figure 1. The measurements are then operated at roughly constant Reynolds number, increasing progressively the pressure in the test-bench.

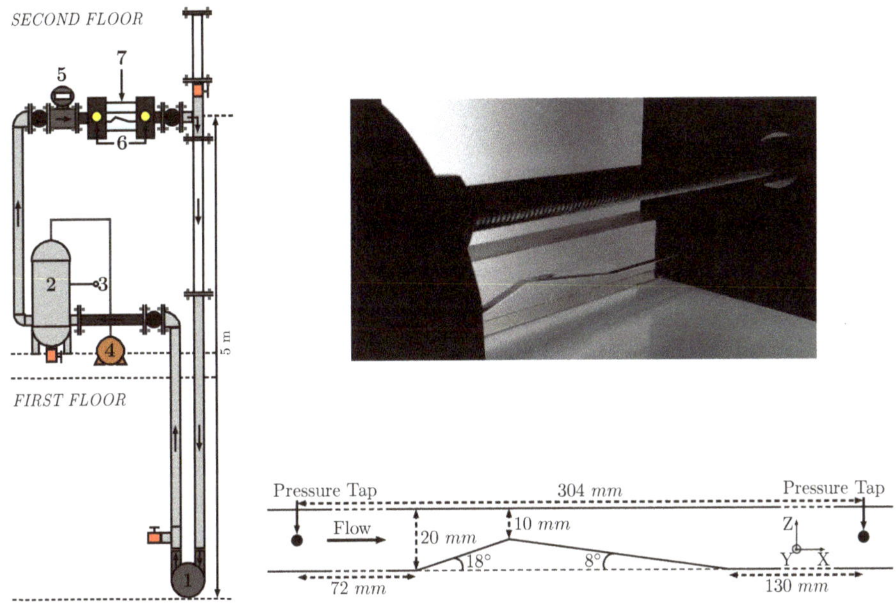

Figure 1. Sketch of the Venturi geometry. The width of the test section is $w = 10$ mm and its height at the inlet section is $h_{in} = 20$ mm. In the following of the article the axes origin is positioned at the Venturi throat edge in the middle of test section along the width. Reprinted with permission from K. Croci, Ph.D. dissertation [32].

2.2. Control Parameters, Flow Visualization and Measured Quantities

The inlet and outlet test section absolute pressure, named P_1 and P_2, are measured with two pressure transducers located respectively at 103 mm upstream and 201 mm downstream of the Venturi throat. The discharge throat velocity V_{in} at S_{in} is computed from flow rate measurements. The viscosity and density have been measured as a function of the temperature [32].

The following dimensionless numbers are defined based on these values: a Reynolds number Re, an inlet and an outlet cavitation numbers σ_1 and σ_2 and a capillary number Ca. The definition of these parameters, together with the ranges that have been explored in the present article and the associated uncertainties, can be found in Table 1. Please note that the outlet cavitation number σ_2 has not been considered in the previous study [28]. It is used in the present article as it seems to be more significant for analyzing the results. We also introduce a dimensionless pressure coefficient K_p. Please note that K_p is simply $K_p = \sigma_1 - \sigma_2$.

Two Optronis CR1000x3 high-speed cameras are used to visualize the cavities. The two cameras allow synchronized pictures from the top and from the side of the test section, with a field of view of 1280×512 pixels. Image sequences of 2 to 3 seconds are recorded at 1500 Hz or 1000 Hz. The flow is illuminated from the bottom and the backside with two continuous white LED plate from Phlox and consequently, the gaseous cavities appear as black features in the images. A typical result obtained at $Re \simeq 1300$ is displayed in Figure 2.

Table 1. Flow parameters, ranges and estimated uncertainties for silicone oil 47V50. The surface tension is $S \simeq 20$ mN·m^{-1} and the vapor pressure is $P_v \simeq 1.3$ Pa according to the oil data file.

Symbol	Parameters	Definition	Range	Unit	Uncertainty
P_1	Inlet pressure		[146; 708]	mbar	±1
P_2	Outlet pressure		[2; 294]	mbar	±1
T	Operating temperature		[18.0; 22.0]	°C	±0.1
V_{in}	Inlet velocity		[2.12; 5.68]	m·s^{-1}	±0.02
ρ	Oil density		[960.6; 962.9]	kg·m^{-3}	±0.2
ν	Oil kinematic viscosity		[52.5; 56.9]	mm^2·s^{-1}	±2%
Ca	Capillary number	$\frac{\rho \nu V_{in}}{S}$	[5.4; 15.1]		⩽3%
K_p	Pressure loss coefficient	$\frac{P_1 - P_2}{\frac{1}{2}\rho V_{in}^2}$	[3.11; 6.50]		⩽4%
Re	Inlet Reynolds number	$\frac{V_{in} h_{in}}{\nu}$	[797; 2056]		⩽3%
σ_1	Inlet cavitation number	$\frac{P_1 - P_v}{\frac{1}{2}\rho V_{in}^2}$	[4.42; 7.61]		⩽2%
σ_2	Outlet cavitation number	$\frac{P_2 - P_v}{\frac{1}{2}\rho V_{in}^2}$	[0.02; 3.50]		⩽2%

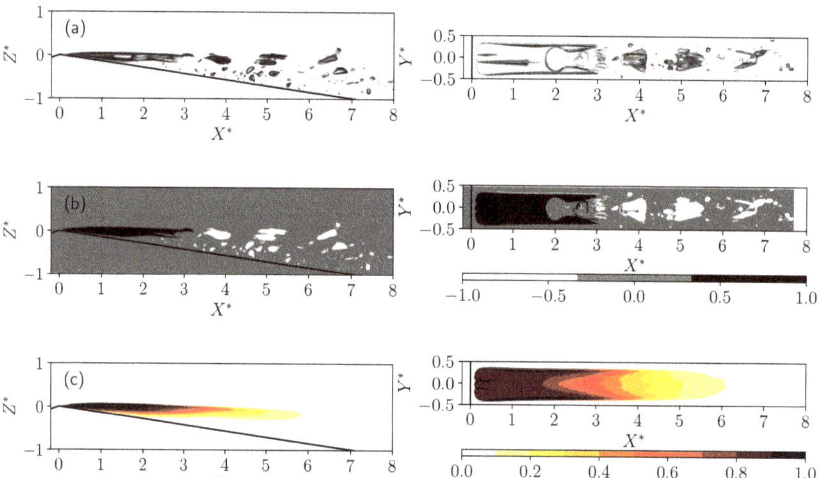

Figure 2. (**a**) Typical side (left column) and top (right column) instantaneous views. The flow is from left to right. $Re \simeq 1300$, $\sigma_1 = 5.10 \pm 0.04$ and $\sigma_2 = 0.70 \pm 0.03$. (**b**) Illustration of the image processing. The treatment is based on binarization, morphological closing and regions labeling (see text for details). The large region close to the throat is displayed in black in the figure, whereas the other small gaseous cavities are in white and the background in pale gray. A binary image (not shown) is generated that only contain the large region close to the throat. (**c**) Time average of 3000 binary images showing the average attached cavity on the side and top views. The colorbar corresponds to the proportion of time when the cavity is present.

The height of the Venturi throat $h_{throat} = 10$ mm is chosen as the reference length scale for the dimensionless coordinates X^*, Y^* and Z^* (see also the bottom right part in Figure 1 for the definition of the coordinate system). The origin of the positions is on top of the wedge, in the symmetry plane of the flow vein. The diverging part of the Venturi section thus extends from $X^* = 0$; $Z^* = 0$ to

$X^* = 1/\tan(8°) \simeq 7.1$; $Z^* = -1$; the two side walls are at $Y^* = \pm 0.5$ and the upper flat wall is at $Z^* = 1$.

Time series of 3000 images are recorded for each experiment. A background reference image I_{ref} is also recorded with the same cameras setting, while the flow is at rest. The instantaneous images I_i are then processed with Python, using the `scipy.ndimage` and `scikit-image` libraries [33] in the following way, based on Morphological Image Processing [34]:

- The images are converted to matrices of floats, the background is subtracted and the result is normalized by the background reference image, giving the ith normalized image of the time series $I_i^{'} = \frac{I_i - I_{ref}}{I_{ref}}$ (see Figure 2a). The gaseous features that are darker than the background have then gray-level values in the range -1 ; 0.
- Image $I_i^{'}$ is then binarized with a threshold of -0.4: $I_i^b = \left(I_i^{'} < -0.4\right)$ which gives the position of the interfaces between liquid and cavities with a dimensionless accuracy of ± 0.02 (see Figure 3b for a gray-level profile on a normalized image).
- The holes that could remain inside the cavities are then filled with an algorithm based on invading the complementary of I_i^b from the outer boundary of the image with binary dilations. Holes are not connected to the boundary and are therefore not invaded. The result is the complementary subset of the invaded region. A filled binary image I_i^f is obtained.
- The features are then labeled, and their properties are measured with the `scikit-image` functions `measure.label()` and `measure.regionprops()`. A filter on the bounding box of the labeled regions is then used to keep the region that is the closest to the Venturi throat, resulting in a binary image containing a closed and filled representation of the cavity that is attached to the throat I_i^a. This process is illustrated in Figure 2b where $2I_i^a - I_i^f$ is displayed.
- The mean I^m and standard deviation I^{rms} of the time series of the filtered images I_i^a are then computed. The mean image for the series at $Re \simeq 1300$, $\sigma_1 = 5.10 \pm 0.04$ and $\sigma_2 = 0.70 \pm 0.03$ is displayed in Figure 2c.

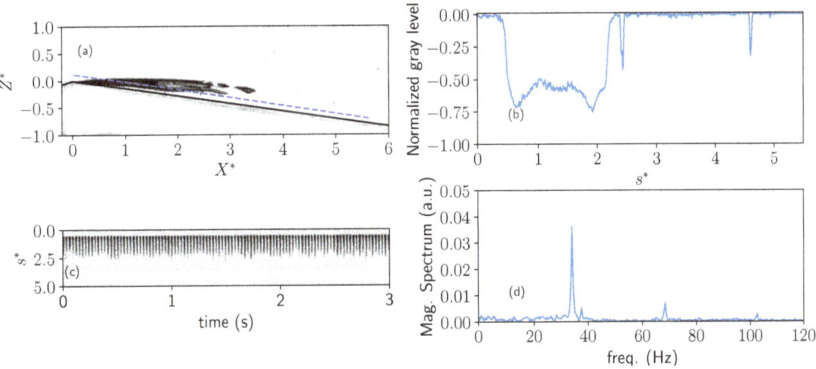

Figure 3. Illustration of image processing for the characterization of the temporal features of the attached cavities. (**a**): Instantaneous normalized side view $I_i^{'}$ at $Re = 1600$, $\sigma_1 = 4.97 \pm 0.04$ and $\sigma_2 = 1.17 \pm 0.03$. (**b**): Gray level along the dashed blue line displayed in (a); s^* is the dimensionless curvilinear abscissa along the line. (**c**): Spatio-temporal diagram along the line; space is from top to bottom and time is from left to right. (**d**): Corresponding temporal magnitude spectrum.

The length of the attached cavities is then measured on the mean images of the side view, using a threshold $I^m > 0.6$. This value corresponds to the region of space where the attached sheet is present on 60% of the time series. For the case illustrated in Figure 2, it gives a length $L^* \simeq 3.5$. Varying the threshold between 0.5 and 0.7—see the colorbar in Figure 2c—leads to a variation of ± 0.2 on the measure of L^* which is a relative variation of $\pm 5\%$.

Spatio-temporal diagrams are also extracted along various lines in the normalized images I_i' in order to characterize the dynamics of the cavities and to measure the frequencies of periodical shrinking or shedding of the cavities. The post-processing is illustrated in Figure 3, where the spatio-temporal diagram is extracted along a line parallel to the Venturi diverging wall, about 1 mm above. The spatio-temporal matrix of gray levels is averaged along the spatial coordinate to give a time signal, from which the magnitude spectrum is computed. The main result which is the frequency of the main peak in the spectrum—if existing—is insensitive to the precise choice of the line: similar diagrams and the same frequency are obtained for horizontal or inclined lines at various heights, providing these lines cross the attached sheet. The best signal-to-noise ratio is obtained for the line displayed in Figure 3.

2.3. Numerical Highlights and Comparison with Experimental Data

The topology of the single-phase flow and its nature as a function of the Reynolds number have been studied numerically [28]. The computation was performed with the Computational Fluid Dynamics code StarCCM+ (version 13.02). It uses a finite volume approach. The computation is three-dimensional; the numerical domain is similar to the experiment with an extrusion of $10h_{throat}$ upstream of the inlet of the test section and an extrusion of $20h_{throat}$ downstream of the outlet of the test section. The boundary conditions are a uniform inlet velocity and a uniform static pressure outlet. The domain is meshed with a structured grid. A grid convergence test was made at $Re = 1500$ with up to 7.2×10^6 cells. The final mesh contains 1.1×10^6 cells, with 51 points in the cross-stream direction and a near-wall cell size of $5 \times 10^{-3}h_{throat}$. The equations that are solved are the incompressible Navier–Stokes equations for a Newtonian fluid with constant properties. The spatial discretization for the convective flux is the second-order upwind scheme. Unsteady simulations with the Euler first-order implicit scheme have been performed. The main results are plotted in Figure 4. The experimental (obtained in single-phase flows) and numerical pressure loss coefficients fairly collapse, which is a clue in favor of the quality of the simulations (Figure 4a).

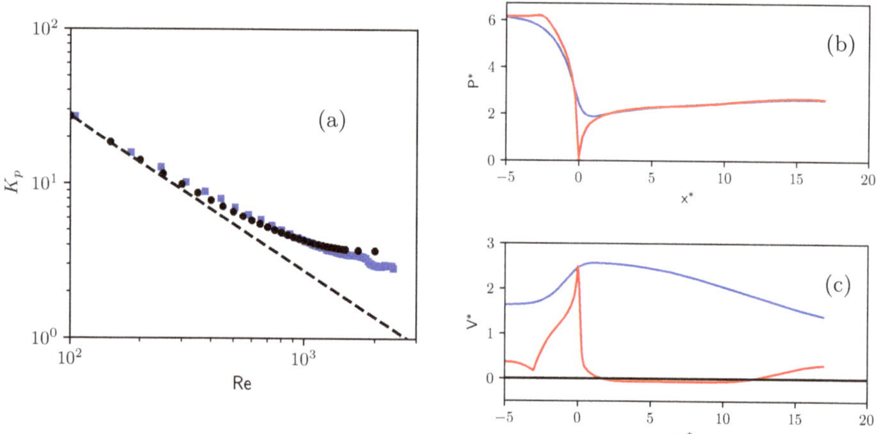

Figure 4. Main results of the single-phase numerical simulations. (**a**) Pressure loss coefficient K_p as a function of the Reynolds number Re. Blue squares represent experimental values while black squares represent numerical simulations. The dashed line is a line of equation $K_p \propto Re^{-1}$. (**b**) Dimensionless pressure profile $P^* = P/\frac{1}{2}\rho V_{in}^2$ with respect to axial coordinate $X^* = X/h_{throat}$. (**c**) Dimensionless velocity profile $V^* = V/V_{in}$ with respect to X^*. For both images (**b**,**c**): $Re = 1200$, $\sigma_1 = 6.57$ and $\sigma_2 = 2.58$; Blue line is along the horizontal line $Z^* = 0.5$ and $Y^* = 0$; Red line is a broken line following the bottom of the vein, 0.5 mm above it at $Y^* = 0$.

The main results of this numerical study are the following:

- The flow remains laminar for $Re \lesssim 2000$. A primary flow separation is present in the wake of the throat, on the divergent wall ($X^* > 0$ and $Z^* < 0$), first along each lateral wall ($Y^* \simeq \pm 0.5$) for $Re \geqslant 350$, and on all along the width for $Re \geqslant 650$. This corresponds to the negative values of the velocity profile in Figure 4c.
- A secondary laminar boundary layer separation emerges at a Reynolds number $Re \geqslant 1100$ at the top wall of the test section ($X^* > 0$ and $Z^* \simeq 1$).
- Finally, the critical cavitation numbers that lead to a minimal absolute pressure equal to the vapor pressure can be estimated numerically with the simulations. For instance, here, for $Re = 1200$, $\sigma_1 = 6.57$ and $\sigma_2 = 2.58$, one can notice that the minimal pressure is close to zero and that this occurs at the throat, on the bottom wall (see Figure 4b). According to the simulation, cavitation of the oil is thus possible.

3. Results

3.1. Comparison between Degassed and Air-Saturated Oil

In the previous study using silicone oil saturated with air at 1 bar (Croci et al. (2019) [28]), three types of gaseous/vaporous structures have been observed. A typical example obtained at $Re = 1810$, $\sigma_1 = 4.81$ and $\sigma_2 = 1.52$ is shown in Figure 5a. The first one is called "bottom tadpoles" (see label (I) in Figure 5a). They attach to the side walls close to the throat, in the region of the primary flow separation that has been numerically demonstrated, and can be distinguished only in the top view. These structures appear for $Re \leqslant 650$ and at inlet pressures well above the estimated critical value for cavitation. They thus might be composed of degassed air. The second type of structure consists of an attached cavity in the middle of the vein, just downstream of the throat (see label (II) in Figure 5a). It has been observed for $Re \leqslant 800$ and their inception fairly corresponds to the critical value of the cavitation number estimated with numerical simulations. This attached cavity thus probably contains vapor. The third type of structures is reminiscent of the tadpoles and has been called "top tadpoles" (see label (III) in Figure 5a): they appear for $Re \geqslant 1000$ and attach to the upper wall in the secondary boundary layer separation. The pressure in these regions is above the vapor pressure and they might also be composed of degassed air.

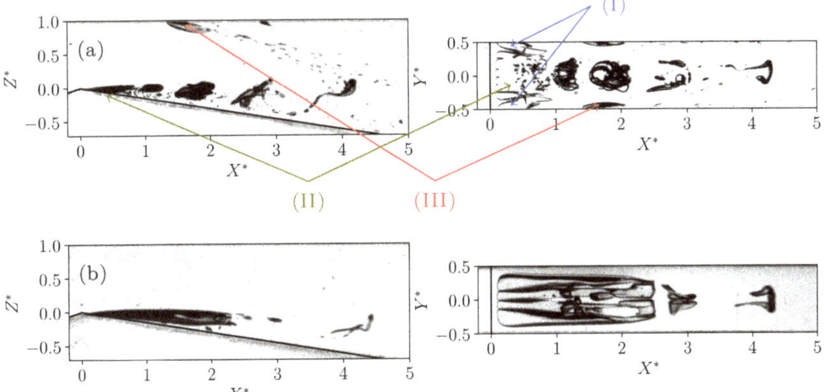

Figure 5. Comparison between two experiments performed at similar Reynolds numbers ($Re \simeq 1800$) and cavitation numbers ($\sigma_2 \simeq 1.5$) with a different degassing protocol. (**a**): Silicone oil is saturated with air at 1 bar before running the experiment. The three types of gaseous structures are spotted with (I) for bottom tadpoles, (II) for attached sheet cavity and (III) for top tadpoles. (**b**): Silicone oil is first degassed at 70 mbars; an attached sheet cavity is only present.

In the present article, with a different protocol of intense degassing at 70 mbars before running the experiments, the bottom and top tadpoles are not observed (see Figure 5b, obtained at $Re = 1810$, $\sigma_1 = 4.91$ and $\sigma_2 = 1.57$). This confirms that the top and bottom tadpoles are linked to degassing phenomena. Only the central attached cavity remains, and these cavities are moreover observed for cavitation numbers lower than the numerical critical value.

The following paragraphs are devoted to the study of the development of these cavities when varying the absolute pressure at constant Reynolds number.

3.2. Developed Cavitation Regimes Identification in Degassed Oil

Approximately one hundred of movies have been recorded at various Reynolds numbers and cavitation numbers. Only the cases with no spurious air bubbles coming from upstream and with cavities larger than 1 mm have been reported in Figure 6.

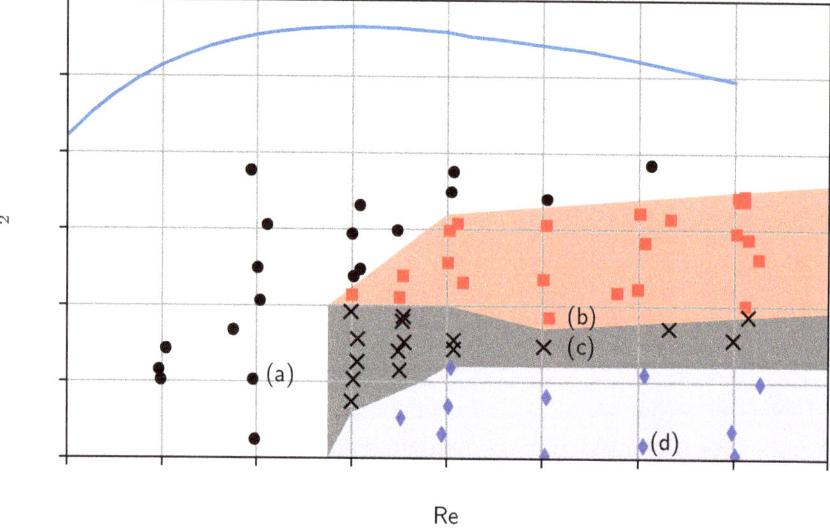

Figure 6. Explored map in the Reynolds—cavitation number ($Re - \sigma_2$) space. Black circles (•) correspond to type-1 steady small cavities, red squares (■) to type-2 cavities with periodic shrinking, crosses ($\times$) to type-3 unsteady cavities with no clear periodic behavior and blue diamonds (♦) to type-4 supercavities with $L^* \geqslant 7.1$. The solid pale blue line corresponds to the critical cavitation numbers obtained with the single-phase numerical simulations. Images of the cases (a) to (d) are displayed in Figure 7.

The data all fall below the numerical inception curves. Four different regimes have been identified. Typical instantaneous side and top views illustrating these four regimes are displayed in Figure 7. The first regime identified with a full circle (•) corresponds to steady cavities. For $Re \leqslant 1000$, even at very small cavitation numbers, only steady cavities of this type are observed. The cavity viewed from side is quite flat and its bottom part is detached from the divergent bottom wall.

For larger Reynolds numbers ($Re \geqslant 1200$), the type-1 steady attached cavities are only observed at high cavitation numbers, corresponding to small cavities. When lowering the cavitation number, a second regime arises: the cavity periodically grows and shrinks, which corresponds to spatio-temporal diagrams looking like the one displayed in Figure 3c. A typical cycle of shrinking and growing is illustrated in Figure 8. This second regime, characterized by a "sawteeth" spatio-temporal diagram is called "type-2 cavities with periodic shrinking" and is displayed with red squares (■).

In this regime, the shape of the cavity is different and varies during the cycle. It consists of an elongated bubble with a narrow tail at maximum length and during the shrinking, one can observe that the tail of the cavity turns into a vertical front that collapses towards the throat. For the case illustrated in Figure 8, the length of the cavity varies between $L^*_{max} \simeq 4$ and $L^*_{min} \simeq 0.2$.

Lowering further the cavitation number, the attached cavity loses this periodic behavior: the spatio-temporal diagrams still show unsteadiness in the flow but with no clear periodic cycle. This third regime is displayed with crosses ($\times$) in Figure 6. The variation of its length corresponds this time to the detachment of delta and hairpin vortices as can be seen in Figure 7c. The cavity in this case keeps a length greater than $L^*_{min} \simeq 1$ and has a maximum length of the order of $L^*_{max} \simeq 6$. The characteristic vertical front at the rear part of the cavity that has been noticed in the previous regime is no longer observed.

Finally, at the lowest cavitation numbers, one can obtain an almost steady cavity that appears to be completely transparent from both top and side view, and whose length is longer than the end of the divergent part of the Venturi profile (i.e., $L^* \geqslant 7.1$). (see Figure 7d). This fourth type is called "supercavities" and is displayed with blue diamonds (♦) in Figure 6.

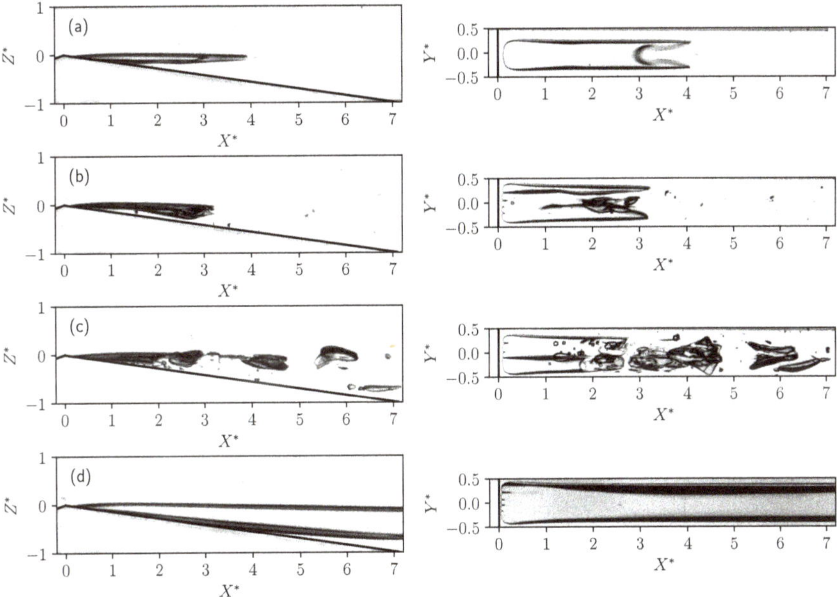

Figure 7. Side and top views in different regimes. (**a**): type-1 steady cavity (•) at $Re = 992$, $\sigma_1 = 5.96$ and $\sigma_2 = 0.51$. (**b**): type-2 cavity with periodic shrinking (■) at $Re = 1614$, $\sigma_1 = 4.87$ and $\sigma_2 = 0.92$. (**c**): type-3 unsteady cavity with no clear periodic behavior ($\times$) at $Re = 1600$, $\sigma_1 = 4.89$ and $\sigma_2 = 0.73$. (**d**): to type-4 supercavity (♦) at $Re = 1810$, $\sigma_1 = 4.66$ and $\sigma_2 = 0.09$.

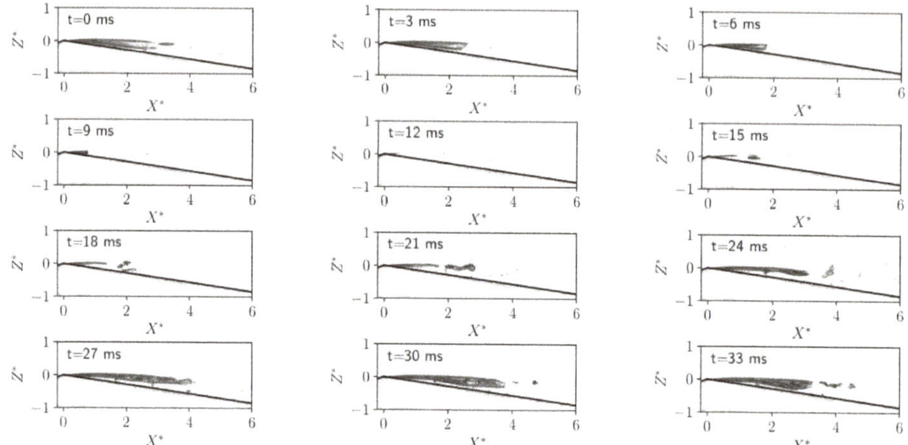

Figure 8. One cycle viewed from side in the type-2 periodic shrinking regime (■) at $Re = 1614$, $\sigma_1 = 4.87$ and $\sigma_2 = 0.92$. The measured frequency of the cycle is 34.6 *Hz*.

3.3. Quantitative Characterization of the Different Regimes and Effects of the Reynolds Number

The type-2 cavities with periodic shrinking are first observed in a very narrow region of the parameter space for $1200 \leqslant Re \leqslant 1350$ and are then present in a wider range of cavitation numbers for $Re \in [1400; 2100]$: they are found to lie in $\sigma_2 \in [0.9; 1.7]$.

The type-4 supercavities have been observed only for one point at very low pressure at $Re \simeq 1300$ ($\sigma_2 \simeq 0.27$). If supercavities exist for $Re = 1200$ it would be at $\sigma_2 \leqslant 0.3$. However, for higher Reynolds numbers, i.e., for $Re \geqslant 1400$ the zone where supercavities are observed progressively extends up to a constant value of the cavitation number $\sigma_2 \leqslant 0.57$.

There thus seems to be a progressive transition in the dynamics of attached sheet cavitation at very low Reynolds number between $Re \simeq 1200$ and $Re \simeq 1400$. Quantitative measurements of the temporal and spatial features of the attached cavities are analyzed and discussed in the next paragraph.

First, the frequency of the periodic shrinking for all the type-2 cavities that have been identified (■ in Figure 6) are plotted as a function of the throat velocity in Figure 9. The two quantities are highly correlated: the frequency seems to be proportional to the throat velocity, following a law $f = 3.83V_{throat}$. A Strouhal number based on the throat velocity and on the throat height as a length scale is $S_t = \frac{fh_{throat}}{V_{throat}}$. The type-2 regime is thus characterized by a constant Strouhal number of the order of $S_t \simeq 0.0026$. Such a low value is reminiscent of the "sheet cavity oscillation" regime described for a turbulent flow in a similar water experiment by Danlos et al. (2014) [10] as opposed to the "cloud cavitation" regime that is characterized by a Strouhal number greater by almost one order of magnitude. Moreover, in that study [10] the sheet cavity oscillation was observed to dominate the dynamics for short cavities and to disappear as the cavity length increases.

The evolution of the average length of the cavities as a function of the outlet cavitation number is plotted in Figure 10. For all Reynolds numbers, the overall trend is an increase of the length with a decrease of the cavitation number, which is not surprising. However, the shape of the curve exhibits a dependence on the Reynolds number. One can indeed notice that the curve becomes steeper as the Reynolds number increases, from $Re \simeq 800$ to $Re \simeq 1300$. At very low Reynolds number the cavity grows relatively slowly with a decrease of the cavitation number and the transition to super-cavitation is progressive at $Re \simeq 1300$. The data obtained for $Re \geqslant 1400$ then follow a different trend, with first a slow growing until $L^* \simeq 1$ and then a faster growing of the cavity. Moreover, one can notice that for $Re \geqslant 1300$, the transition between the type-2 regime (■) and the type-3 regime ($\times$) corresponds to an average length $L^* \simeq 2$, around $\sigma_2 \simeq 1$.

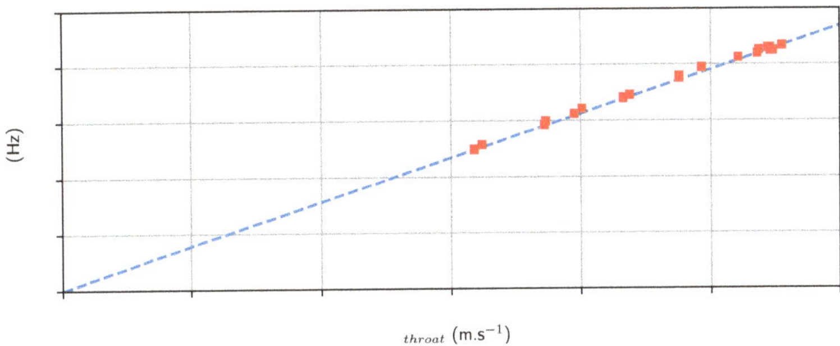

Figure 9. Measured frequency as a function of the throat velocity for type-2 regime. The dashed line is a linear fit of equation $f = 3.83V_{throat}$.

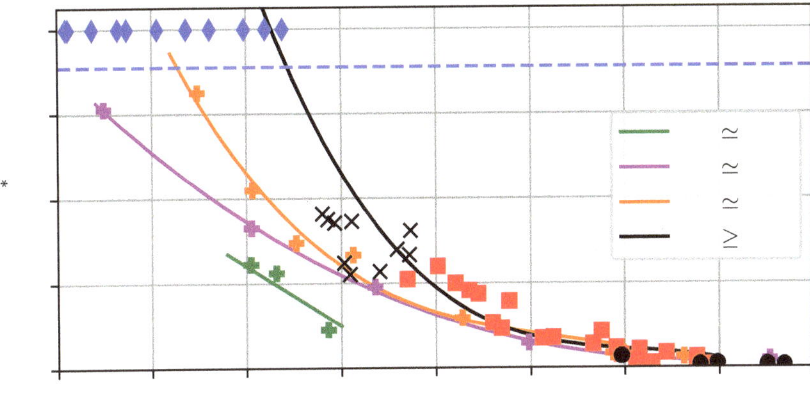

Figure 10. Dimensionless average length of the attached cavity L^* as a function of the outlet cavitation number σ_2. For $Re \geqslant 1300$, the symbols refer to the regimes identified in Figure 6: type-1 steady cavity (•), type-2 cavity with periodic shrinking (■), type-3 unsteady cavity with no clear periodic behavior (×) and type-4 supercavity (♦). The lines are a guide to the eye based on spline interpolation of the points at $Re = 800$, $Re = 1000$, $Re = 1200$ and of all the points for $Re \geqslant 1400$.

The dimensionless pressure drop K_p in the experiment without cavitation, i.e., at $\sigma_{1,2} >> 1$, is plotted in Figure 4a as a function of the Reynolds number. Let us recall that $K_p = \sigma_1 - \sigma_2$. As the cavitation develops while decreasing the ambient pressure at constant discharge velocity, the presence of gaseous structures of increasing volume may change the pressure drop. It might thus be interesting to explore the variation of both the inlet and outlet pressure, i.e., in a dimensionless form of both σ_1 and σ_2. The experiments are plotted in this parameter space in Figure 11. The points at constant Reynolds number are plotted with the same symbol.

The development of cavitation associated with a decrease of the ambient pressure corresponds to a path from top right to bottom left in this parameter space. For the seven Reynolds numbers that are displayed in the Figure 11, one can first notice that the points follow a similar trend: the two pressures are decreasing together. Moreover, one can distinguish two main different shapes in the curves. On the one hand, for $Re \simeq 1000$ and $Re \simeq 1200$, the points align quite on a straight path. On the other hand, for $Re \geqslant 1300$, one can notice some kind of break in the slope of the curve around $\sigma_2 \simeq 1$. For $\sigma_2 \geqslant 1$, the slope is of the order of magnitude of the former (roughly -2), whereas for $\sigma_2 \leqslant 1$, the slope is steeper, of the order of -10: a very small decrease in the inlet pressure causes a dramatic decrease of the outlet pressure corresponding to the fast increase of the cavity length observed in Figure 10.

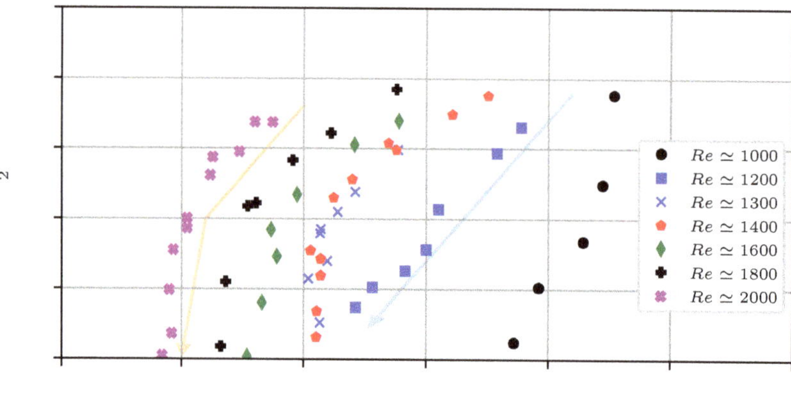

Figure 11. Explored map in the inlet cavitation number—outlet cavitation number ($\sigma_1 - \sigma_2$). The different symbols stand for different Reynolds numbers. The arrows are guides to the eye.

These observations support the idea of a transition in the cavitating flow that takes place around $Re \simeq 1300$.

Looking at the attached cavities from top (see the right column in Figures 2, 5 and 7), one can notice that the morphology of the cavity also exhibits a dependence in the transverse direction Y^*. At very low Reynolds number (see Figure 7a) the interface at the leading edge presents a single smooth and convex cap. At larger Reynolds numbers (see Figure 7d) this interface presents several indentations. These structures resemble the "divots" that have been discussed for instance in Ref. [35] and that are attributed to a local breakdown of the two-dimensionality of the laminar boundary layer. We report in Figure 12 the number of indentations that are observed on the cavities of an average length of $L^* \geqslant 4$. The first indentation is observed at $Re \simeq 1200$ and then their number increases almost linearly with the Reynolds number.

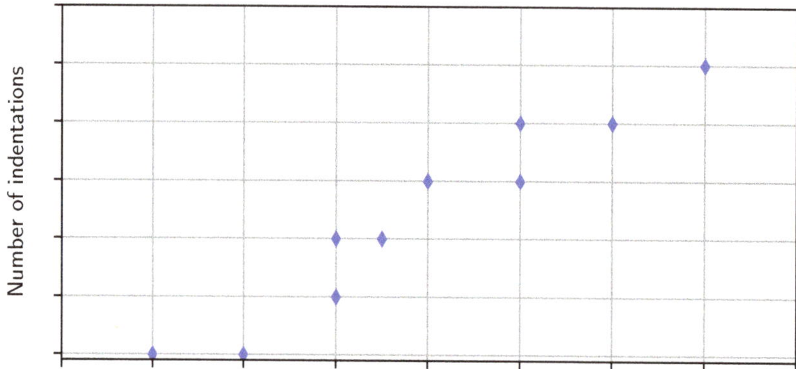

Figure 12. Number of transverse indentations at the leading edge of the developed attached cavities as a function of the Reynolds number.

4. Conclusions

In the present article, the development of attached cavities under cavitating conditions in a highly degassed silicone oil has been explored, at Reynolds numbers in the range $Re \in [800; 2000]$. The regime map in terms of Reynolds and cavitation numbers have been established. Four different regimes have been observed: type-1 small cavities that are almost steady, type-2 periodically shrinking cavities of intermediate sizes, type-3 vortex bursting with no clear period and type-4 supercavities. For $Re \leqslant 1000$, the background flow is steady and laminar and the cavitating flow gives rise to steady attached cavities whose length slowly increases with a decrease of the cavitation number. The first footprint of unsteadiness in the cavitating flow consists of the type-2 periodic regime. It appears for a sufficiently great Reynolds number ($Re \geqslant 1200$), first in a very narrow range of cavitation numbers and then in a wider range as $Re \geqslant 1400$. The frequency is almost proportional to the velocity of the upstream flow and gives a constant Strouhal number of very small value. More modal analysis with POD (Proper Orthogonal Decomposition) or DMD (Dynamic Mode Decomposition) is in progress to characterize better this regime. These experiments will require higher temporal resolution. The analysis of the length of the cavity, of the transition to super-cavitation and of the evolution of the pressure drop are consistent with a smooth transition in the cavitating flow for $1200 \geqslant Re \geqslant 1400$, which corresponds to the range where the single-phase (non-cavitating) flow is transitional. It would be interesting to extend further the range of Reynolds numbers, using an oil of lower viscosity, in order to give a lower bound in Reynolds number for the periodical cloud shedding regime that is known to take place in this geometry in fully developed turbulent flows.

Author Contributions: Conceptualization, F.R., F.B., A.D. and S.K.; methodology, F.R.; software, F.R.; validation, F.R., F.B., A.D. and S.K.; formal analysis, F.R.; investigation, F.R. and A.D; resources, K.C. and C.S.; data curation, F.R. and A.D.; writing–original draft preparation, F.R.; writing–review and editing, F.R.; visualization, F.R.; supervision, F.B. and S.K.; project administration, F.R.; funding acquisition, F.B. and S.K. All authors have read and agreed to the published version of the manuscript.

Funding: This research received no external funding.

Acknowledgments: The authors acknowledge Marc Joulin and Jocelyn Mistigres for technical support, and Michaël Pereira for fruitful discussion.

Conflicts of Interest: The authors declare no conflict of interest.

References

1. Brennen, C.E. *Fundamentals of Multiphase Flows*; Cambridge University Press: Cambridge, England, 2005.
2. Franc, J.P.; Michel, J.M. *Fundamentals of Cavitation*; Kluwer Academic Publishers: Dordrecht, The Netherlands, 2006.
3. Parkin, B.R.; Kermeen, R.W. *Incipient Cavitation and Boundary Layer Interaction on a Streamlined Body*; Rep. E-35.2; California Institute of Technology: Pasadena, CA, USA, 1953.
4. Arakeri, V.H. Viscous effects on the position of cavitation separation from smooth bodies. *J. Fluid Mech.* **1975**, *68*, 779–799. [CrossRef]
5. Kuiper, G. Cavitation Inception on Ship Propeller Models. Ph.D. Thesis, Delft University of Technology, Delft, The Netherlands, 1981.
6. Franc, J.P.; Michel, J.M. Attached cavitation and the boundary layer: Experimental investigation and numerical treatment. *J. Fluid Mech.* **1985**, *154*, 63–90. [CrossRef]
7. Guennoun, M.F. Étude Physique de L'apparition et du Développement de la Cavitation sur une aube Isolée. Ph.D. Thesis, EPFL Lausanne, Lausanne, Switzerland, 2006.
8. van Rijsbergen, M. A review of sheet cavitation inception mechanisms. In Proceedings of the 16th International Symposium on Transport Phenomena and Dynamics of Rotating Machinery, Honolulu, HI, USA, 10–15 April 2016. Available online: https://hal.archives-ouvertes.fr/hal-01890067/ (accessed on 10 October 2020).
9. Brandner, P.; Walker, G.J.; Niekamp, P.N.; Anderson, B. An experimental investigation of cloud cavitation about a sphere. *J. Fluid Mech.* **2010**, *656*, 147–176. [CrossRef]

10. Danlos, A.; Ravelet, F.; Coutier-Delgosha, O.; Bakir, F. Cavitation regime detection through Proper Orthogonal Decomposition: Dynamics analysis of the sheet cavity on a grooved convergent divergent nozzle. *Int. J. Heat Fluid Flow* **2014**, *47*, 9–20. [CrossRef]
11. Pelz, P.F.; Keil, T.; Groß, T.F. The transition from sheet to cloud cavitation. *J. Fluid Mech.* **2017**, *817*, 439–454. [CrossRef]
12. Šarc, A.; Kosel, J.; Stopar, D.; Oder, M.; Dular, M. Removal of bacteria Legionella pneumophila, Escherichia coli, and Bacillus subtilis by (super)cavitation. *Ultrason. Sonochem.* **2018**, *42*, 228–236. [CrossRef]
13. Kosel, J.; Šuštaršič, M.; Petkovšek, M.; Zupanc, M.; Sežun, M.; Dular, M. Application of (super)cavitation for the recycling of process waters in paper producing industry. *Ultrason. Sonochem.* **2020**, *64*, 10500. [CrossRef]
14. Tsujimoto, Y.; Kenjiro, K.K.; Brennen, C.E. Unified treatment of flow instabilities of turbomachines. *J. Propuls. Power* **2001**, *17*, 636–643. [CrossRef]
15. Campos-Amezcua, R.; Bakir, F.; Campos-Amezcua, A.; Khelladi, S.; Palacios-Gallegos, M.; Rey, R. Numerical analysis of unsteady cavitating flow in an axial inducer. *Appl. Therm. Eng.* **2015**, *75*, 1302–1310. [CrossRef]
16. Callenaere, M.; Franc, J.P.; Michel, J.M.; Riondet, M. The cavitation instability induced by the development of a re-entrant jet. *J. Fluid Mech.* **2001**, *444*, 223–256. [CrossRef]
17. Ganesh, H.; Mäkiharju, S.A.; Ceccio, S.L. Bubbly shock propagation as a mechanism for sheet-to-cloud transition of partial cavities. *Phys. Fluids* **2016**, *802*, 37–78. [CrossRef]
18. Croci, K.; Tomov, P.; Ravelet, F.; Danlos, A.; Khelladi, S.; Robinet, J.C. Investigation of two mechanisms governing cloud cavitation shedding: experimental study and numerical highlight. In Proceedings of the ASME International Mechanical Engineering Congress and Exposition, Phoenix, AZ, USA, 11–17 November 2016; Volume 7.
19. Wu, J.; Ganesh, H.; Ceccio, S. Multimodal partial cavity shedding on a two-dimensional hydrofoil and its relation to the presence of bubbly shocks. *Exp. Fluids* **2019**, *60*, 66. [CrossRef]
20. Brunhart, M.; Soteriou, C.; Gavaises, M.; Karathanassis, I.; Koukouvinis, P.; Jahangir, S.; Poelma, C. Investigation of cavitation and vapor shedding mechanisms in a Venturi nozzle. *Phys. Fluids* **2020**, *32*, 083306. [CrossRef]
21. Petkovšek, M.; Hocevar, M.; Dular, M. Visualization and measurements of shock waves in cavitating flow. *Exp. Therm. Fluid Sci.* **2020**, *119*, 110215. [CrossRef]
22. Trummler, T.; Schmidt, S.; Adams, N. Investigation of condensation shocks and re-entrant jet dynamics in a cavitating nozzle flow by Large-Eddy Simulation. *Int. J. Multiph. Flow* **2020**, *125*, 103215. [CrossRef]
23. Ishihara, T.; Ouchi, M.; Kobayashi, T.; Tamura, N. An experimental study on cavitation in unsteady oil flow. *Bull. JSME* **1979**, *22*, 1099–1106. [CrossRef]
24. Washio, S.; Kikui, S.; Takahashi, S. Nucleation and subsequent cavitation in a hydraulic oil poppet valve. *Proc. Inst. Mech. Eng. Part C J. Mech. Eng. Sci.* **2009**, *224*, 947–958. [CrossRef]
25. Peters, F.; Honza, R. A benchmark experiment on gas cavitation. *Exp. Fluids* **2014**, *55*, 1786. [CrossRef]
26. Groß, T.F.; Pelz, P.F. Diffusion-driven nucleation from surface nuclei in hydrodynamic cavitation. *J. Fluid Mech.* **2017**, *830*, 138–164. [CrossRef]
27. Croci, K.; Ravelet, F.; Robinet, J.C.; Danlos, A. Experimental Study of Cavitation in Laminar Flow. In Proceedings of the 10th International Symposium on Cavitation (CAV2018), Baltimore, MD, USA, 14–16 May 2018. [CrossRef]
28. Croci, K.; Ravelet, F.; Danlos, A.; Robinet, J.C.; Barast, L. Attached cavitation in laminar separations within a transition to unsteadiness. *Phys. Fluids* **2019**, *31*, 063605. [CrossRef]
29. Ding, C.; Fan, Y. Measurement of diffusion coefficients of air in silicone oil and in hydraulic oil. *Chin. J. Chem. Eng.* **2011**, *19*, 205–211. [CrossRef]
30. Li, B.; Gu, Y.; Chen, M. An experimental study on the cavitation of water with dissolved gases. *Exp. Fluids* **2017**, *58*, 164. [CrossRef]
31. Amini, A.; Reclari, M.; Sano, T.; Farhat, M. Effect of gas content on tip vortex cavitation. In Proceedings of the Symposium on Cavitation, Baltimore, MD, USA, 14–16 May 2018.
32. Croci, K. Experimental Study of Multiphase Flows within a Separated Laminar Boundary Layer. Ph.D. Thesis, Ecole Nationale Supérieure D'arts et métiers—ENSAM, Paris, France, 2018.

33. van der Walt, S.; Schönberger, J.L.; Nunez-Iglesias, J.; Boulogne, F.; Warner, J.D.; Yager, N.; Gouillart, E.; Yu, T.; the scikit-image contributors. scikit-image: Image processing in Python. *PeerJ* **2014**, *2*, e453. [CrossRef] [PubMed]
34. Soille, P. *Morphological Image Analysis*; Springer: Berlin, Germany, 2004.
35. Tassin Leger, A.; Bernal, L.P.; Ceccio, S.L. Examination of the flow near the leading edge of attached cavitation. Part 2. Incipient breakdown of two-dimensional and axisymmetric cavities. *J. Fluid Mech.* **1998**, *376*, 91–113. [CrossRef]

Publisher's Note: MDPI stays neutral with regard to jurisdictional claims in published maps and institutional affiliations.

Article

Compressible Two-Phase Viscous Flow Investigations of Cavitation Dynamics for the ITTC Standard Cavitator

Ville M. Viitanen [1,*], Tuomas Sipilä [2], Antonio Sánchez-Caja [1] and Timo Siikonen [3]

1 VTT Technical Research Centre of Finland Ltd., 02150 Espoo, Finland; antonio.sanchez@vtt.fi
2 ABB Marine and Ports, 00980 Helsinki, Finland; tuomas.sipila@fi.abb.com
3 Department of Mechanical Engineering, Aalto University, 02150 Espoo, Finland; timo.siikonen@aalto.fi
* Correspondence: ville.viitanen@vtt.fi

Received: 4 September 2020; Accepted: 28 September 2020; Published: 7 October 2020

Featured Application: The present method can be applied to study characteristics of hydrodynamic cavitation and its consequences, such as performance variations and noise, in propellers and rotating turbomachinery. Compressible formulation of the numerical method ensures a more precise treatment of the multiphase flow dynamics, and can reveal some interesting physical phenomena due to the cavitation. Accurate prediction of the multiphase flow is a requisite for optimal, efficient and environmentally friendly design and operation of such devices.

Abstract: In this paper, the ITTC Standard Cavitator is numerically investigated in a cavitation tunnel. Simulations at different cavitation numbers are compared against experiments conducted in the cavitation tunnel of SVA Potsdam. The focus is placed on the numerical prediction of sheet-cavitation dynamics and the analysis of transient phenomena. A compressible two-phase flow model is used for the flow solution, and two turbulence closures are employed: a two-equation unsteady RANS model, and a hybrid RANS/LES model. A homogeneous mixture model is used for the two phases. Detailed analysis of the cavitation shedding mechanism confirms that the dynamics of the sheet cavitation are dictated by the re-entrant jet. The break-off cycle is relatively periodic in both investigated cases with approximately constant shedding frequency. The CFD predicted sheet-cavitation shedding frequencies can be observed also in the acoustic measurements. The Strouhal numbers lie within the usual ranges reported in the literature for sheet-cavitation shedding. We furthermore demonstrate that the vortical flow structures can in certain cases develop striking cavitating toroidal vortices, as well as pressure wave fronts associated with a cavity cloud collapse event. To our knowledge, our numerical analyses are the first reported for the ITTC standard cavitator.

Keywords: hydrodynamic cavitation; compressible two-phase flow; turbulence modelling; system instabilities

1. Introduction

Cavitation, when it occurs, is a cause of many detrimental effects on marine propellers and turbomachinery. Apart from performance degradation, it may trigger erosion, cause vibration problems, or result in high-intensity noise emission on a broad frequency range. Different types of cavitation involve complex features such as bubbly flow, bubble coalescence, nucleation, free-surface instability, and turbulence. These features do not necessarily obey a single but multiple length and time scales, which renders their reliable prediction, either experimentally or numerically, a major undertaking. For instance, when investigating the performance of marine propellers, depending on their operating conditions, many different forms of cavitation can take place simultaneously, such as steady attached sheet cavitation near the blade's leading edge, cloud cavitation in the propeller's wake, and vortex cavitation originating from the blade tips and the propeller hub.

Even in the case of steady uniform inflow conditions, we can observe unsteady behavior of the cavitation. In particular, the shedding behavior of sheet cavitation (Figure 1) and its subsequent transformation to cloudy and bubbly cavities are important in the generation of the aforementioned erosion and noise issues, since upon collapsing these cavitation types can yield strong pressure fluctuations and even shock waves in the bubbly mixture or in the liquid phase [1,2]. A sketch of the shedding cycle of two-dimensional sheet cavitation is shown in Figure 1. The first frames depict the growth of the attached sheet cavity following the break-off of the previous sheet. The growth will cease after a certain length of the sheet cavity, and a strong re-entrant jet is formed at the closure of the sheet cavity, as depicted in Frame 3. The following frame shows the impingement of the cavity interface by the jet, as the jet reaches the cavity near the leading edge. The rear part of the sheet cavity breaks off and reduces to a bubble cloud, which is convected downstream as shown in Frame 5. As the cloud travels downstream, parts of it may undergo a violent collapsing stage, possibly causing noise and material erosion. Meanwhile, as seen in Frame 6, the newly formed sheet cavity on the foil behind the LE will begin to grow, initiating a subsequent break-off cycle.

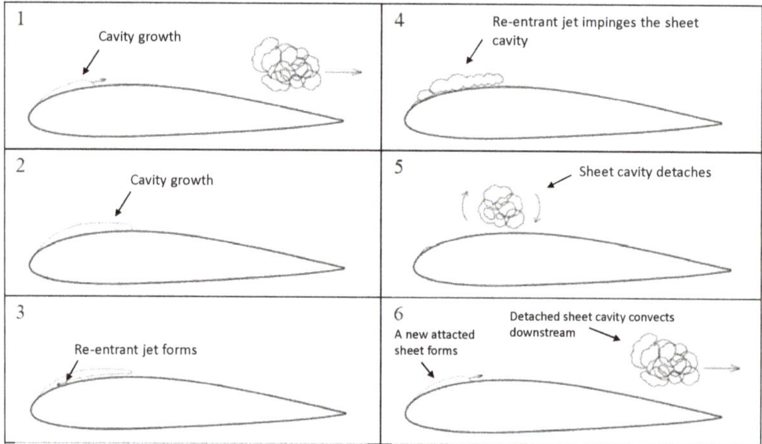

Figure 1. Sketch of the shedding mechanism of sheet cavitation. The figure illustrates one break-off cycle, adapted from [3].

The dynamics of the sheet cavitation are mainly determined by the re-entrant jet, although a bubbly shock wave can in certain conditions affect this development [1]. If the sheet cavity is thick, the break-off cycle is relatively periodic with approximately constant shedding frequency. In these cases, a Strouhal number $St = fl_{ref}/U_{ref}$ can be determined. The length scale of the Strouhal number is chosen as the maximum length of the cavity. This can be argued to be a typical length scale of the cavity since the shedding frequency depends on the distance the re-entrant jet has to travel. The choice of the velocity scale is somewhat problematic, but can be taken as the free-stream velocity $U_{ref} = U_\infty$, or based on an approximation of the re-entrant jet velocity $U_{ref} = U_\infty\sqrt{1+\sigma}$, where σ is the cavitation number, which acknowledges the presence of a cavity structure in a better way than U_∞ alone [4]. The frequency, f, is based on global shedding of the cavity clouds, determined with a visual inspection of the cavity shedding periodicity, with the help of time histories of the measured lift and drag forces. Typically, the Strouhal numbers for sheet-cavitation shedding vary as 0.20 . . . 0.50, depending on the investigated conditions and the definition for the reference quantities [4–7].

In frames 5–6 of Figure 1 we observe that the detached cavity clouds convect downstream toward a higher-pressure region. Upon reaching the regions of sufficiently higher ambient pressure, the detached cavity clouds likely undergo a violent collapsing stage. Collapse events of cavitation structures behind a ship's propeller, for instance, can lead to very high-pressure peaks, and such peaks can exhibit very rapid temporal variation. As noted by Matusiak [8], a primary source of high-frequency noise is the collapse of the free cavitation bubbles. The shedding behavior is mainly governed by the development of the re-entrant jet, vortical structures and turbulent flow near the foil, i.e., the unsteady dynamics of especially the liquid phase. As Reynolds-averaged Navier–Stokes (RANS) equations are mainly derived for steady single-phase flows, complex unsteady phenomena may be severely under-predicted when using an unsuitable turbulence-modelling approach. High-fidelity turbulence modelling is typically needed to capture a detailed flow field for the prediction of the dynamics of cavitation and its consequences such as erosion and noise [9–12]. Therefore in this study, we employ two different turbulence closures: an unsteady RANS-based SST $k-\omega$ turbulence model and a hybrid RANS/LES (large eddy simulation) closure called the delayed-detached eddy simulation (DDES). We analyze the two-phase flow using a compressible multiphase model. Cavitating problems have been typically treated with incompressible flow solvers [13–17] (especially in the case of marine propellers), and recently either isothermal compressible or fully compressible flow formulations have been used as well [18–21]. If we aim for a precise treatment of hydrodynamic cavitation in its various forms, a compressible multiphase flow model is a requisite for reliable numerical predictions [18]. Peculiarly, the sound speed in a mixture can drop to a fraction of that in the liquid and even the vapor phases. Prediction of the correct acoustic signal speeds calls for a complete flow model.

In this paper, the dynamics of the unsteady sheet cavitation is studied in detail. We employ a compressible two-phase flow model to account for the cavitation phenomena, and the computational case is the ITTC standard cavitator [22]. The results are compared to the model tests carried out in a cavitation tunnel. The flow simulations have been conducted with the general purpose CFD solver FINFLO. The cavitation tunnel tests, which include hydrophone measurements of the underwater noise, have been conducted at SVA Potsdam. We have carried out simulations of two different cases. The inflow velocity for both cases was 8 m/s, and the cavitation numbers were $\sigma = [0.53, 1.27]$. To the knowledge of the authors, our results include the first reported numerical studies of the ITTC standard cavitator.

This paper is organized as follows. Section 2 describes the flow solver used, as well as the turbulence and multiphase flow modelling. In Section 3 we describe the test case. Section 4 presents the results in terms of cavitation observation and its dynamics, and their comparison to the experimental findings. Lastly, we present conclusions and suggestions for future work in Section 5.

2. Flow Solution

2.1. Governing Equations

Our flow model is based on a compressible form of the RANS equations where an assumption is made on the homogeneity of the fluid mixture among the different phases involved in the computation [23,24]. Including the energy equations for the solution allows prediction of the correct acoustic signal speeds. The governing continuity and momentum equations are

$$\begin{aligned} \frac{\partial \alpha_k \rho_k}{\partial t} + \nabla \cdot \alpha_k \rho_k \mathbf{V} &= \Gamma_k, \\ \frac{\partial \rho \mathbf{V}}{\partial t} + \nabla \cdot \rho \mathbf{V}\mathbf{V} + \nabla p &= \nabla \cdot \boldsymbol{\tau}_{ij} + \rho \mathbf{g}, \end{aligned} \tag{1}$$

where p is the pressure, $\mathbf{V}$ the absolute velocity in a global non-rotating coordinate system, $\boldsymbol{\tau}_{ij}$ the stress tensor, α_k a void (volume) fraction of phase k, ρ_k the density, t the time, Γ_k the mass-transfer term, and $\mathbf{g}$ the gravity vector. The void fraction is defined as $\alpha_k = \mathscr{V}_k/\mathscr{V}$, where $\mathscr{V}_k$ denotes the volume occupied by phase k of the total volume, $\mathscr{V}$. For the mass transfer, $\sum_k \Gamma_k = 0$ holds, and consequently only a single mass-transfer term is needed. The energy equations for phase $k = g$ or l, the indices referring to gas (g) and liquid (l) phases, are written as

$$\frac{\partial \alpha_k \rho_k (e_k + \frac{V^2}{2})}{\partial t} + \nabla \cdot \alpha_k \rho_k (e_k + \frac{V^2}{2})\mathbf{V} =$$
$$-\nabla \cdot \alpha_k \mathbf{q}_k + \nabla \cdot \alpha_k \boldsymbol{\tau}_{ij} \cdot \mathbf{V} + q_{ik} + \Gamma_k (h_{k\mathrm{sat}} + \frac{V^2}{2}) + \alpha_k \rho_k \mathbf{g} \cdot \mathbf{V}. \quad (2)$$

Here, e_k is the specific internal energy, $\mathbf{q}_k$ the heat flux, q_{ik} interfacial heat transfer from the interface to phase k, and $h_{k\mathrm{sat}}$ saturation enthalpy. The interfacial heat transfer coefficients are based on the mass-transfer model by assuming a saturated temperature for the gas phase, and the heat transfer is then determined by the mass-transfer model. In the case of a homogeneous flow, the resulting sound speed c for a two-phase mixture is obtained from the eigenvalues of the Jacobian of the flux vector as

$$\frac{1}{\rho c^2} = \frac{\alpha}{\rho_g c_g^2} + \frac{1-\alpha}{\rho_l c_l^2} \text{ and } \frac{1}{c_k^2} = \frac{\partial \rho_k}{\partial p} + \frac{1}{\rho_k}\frac{\partial \rho_k}{\partial h_k}. \quad (3)$$

In the expressions above, h_k denotes the enthalpy of phase k.

The momentum and total continuity equations in the homogeneous model do not change, except for the material properties like density and viscosity. They are calculated for the mixture as a weighted sum by the phasic volume fractions. The turbulence effects are those of single-phase models for the mixture with material properties derived from the pressure and temperatures of the individual phases. The determination of the material properties is described in [25].

2.2. Turbulence Modelling

Nominally, a Reynolds-averaged form of the Navier–Stokes equations is used, and the DDES approach [26] that combines RANS and LES is also applied in the same form. In this study, the calculations are performed up to the wall both in the model- and full-scale simulations, which avoids the use of wall functions. The height of the first cell was adjusted such that the non-dimensional wall distance $y^+ = \rho u_\tau y/\mu \lesssim 1$ for the first cell, with $u_\tau = \sqrt{\tau_w/\rho}$ being the friction velocity, τ_w the wall shear stress, and y is the normal distance from the solid surface to the center point of the cell next to the surface.

The base turbulence closure applied in the present calculations is the SST $k-\omega$ model [27]. That is a zonal model, referring to the formulation where the $k-\omega$ equations are solved only inside the boundary layer, and the standard $k-\epsilon$ equations, transformed to the ω-formulation, are solved away from the walls. The DDES is also based on the SST model [28], and described in [12]. DDES is a slightly modified version of the detached-eddy simulation (DES). Both are hybrid RANS/LES models, and function as an LES subgrid-scale model in regions where the local turbulent phenomena are of greater size than the local grid spacing, and reduce to a RANS model in regions where the largest turbulent fluctuations are of a smaller size than the local grid spacing [29]. A time-accurate solution is made to resolve turbulent fluctuations.

2.3. Mass and Energy Transfer

Several mass-transfer models have been suggested for cavitating problems. Usually, the mass-transfer rate is proportional to a pressure difference from a saturated state or to a square root of that. In this study, we employ a mass-transfer model that is similar to that of Choi and Merkle [30], described in [24,31]. The empirical parameters of the cavitation model are calibrated by Sipilä [3]. The saturation pressure is based on the free-stream temperature, and the gas phase is assumed to be saturated.

2.4. Solution Algorithm

The governing equations are discretized in a finite volume manner with viscous fluxes and pressure terms centrally differenced. For the convective part, a third-order upwind-biased monotonic upstream-centered scheme for conservation laws (MUSCL) interpolation is used for the variables on the cell surfaces. A compressive flux limiter that has been shown to improve the predicted tip vortex cavitation pattern [24] is employed for the void fraction in the present study. For the time derivatives, a second-order three-level fully implicit method is used.

The solution method is a segregated pressure-based algorithm where the momentum equations are solved first, and then a pressure-velocity correction is made. The basic idea in the solution of all equations is that the mass balance is not forced at every iteration cycle, but rather the effect of the mass error is subtracted from the linearized conservation equations. A pressure correction equation was derived from the continuity equation, Equation (1), linked with the linearized momentum equation. The velocity-pressure coupling is based on the corresponding algorithm for a single-phase flow [23], and is described in more detail in [12,24]. A physical time step of $\Delta t = 2.5 \times 10^{-5}$ s is used in the simulations, which resolves the cavitation dynamics cycles in the investigated cases with about 250 ($\sigma = 1.27$) and 1200 ($\sigma = 0.53$) steps depending on the conditions. Approximately 100 inner iterations were required within each physical time step.

3. Test Case

A schematic view of the ITTC standard cavitator is shown in Figure 2, and the cavitator inside the cavitation tunnel is seen in Figure 3. The rectangular cross-section of the tunnel has the dimensions of 600×600 mm^2. Two cases with different cavitation numbers were investigated. The inflow velocity for both cases was $U_\infty = 8$ m/s, and the cavitation numbers were

$$\sigma = \frac{p - p_{sat}}{\frac{1}{2}\rho_\infty U_\infty^2} = [0.53, 1.27] \, .$$

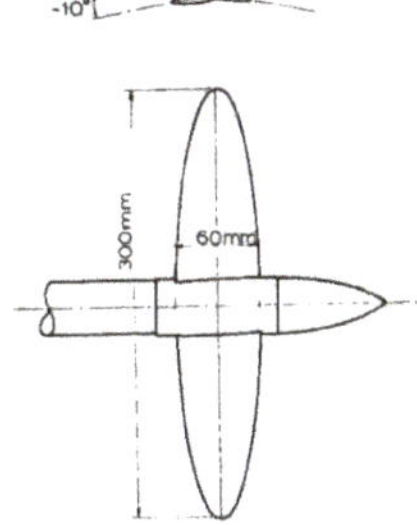

Figure 2. Schematic view of the cavitator.

Three grid densities were used in the calculations. The details are summarized in Table 1. The structured grid has an O-O topology around the foils. The finest grid level consists of roughly 7.3 million cells in 80 grid blocks. The grid resolution around the leading and trailing edges is reasonable, and there are about 15 cells around the leading-edge radius. Due to the O-O topology, the same resolution is applied around the blade tip and the trailing edge as well. In this work, we emphasized the resolution of the sheet cavitation on the wings, and for this reason no clustering of cells was targeted for the cavitating tip vortex, for instance. The grid is refined normal to the viscous surfaces such that $y^+ \approx 1$ at the finest grid level. A detail of the grid is shown in Figure 4, and the grid resolution on the surface of the foil is shown in Figure 5.

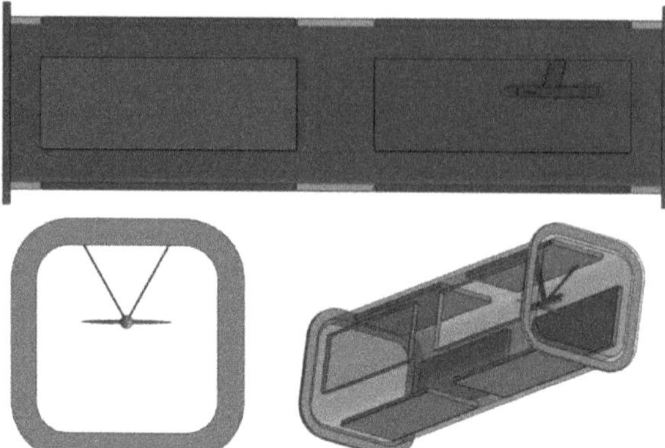

Figure 3. Placement of the cavitator inside the cavitation tunnel.

Table 1. Summary of the details of the grid used in the calculations for the three grid levels. The number of surface cells are given for one side of the foil.

	Fine Grid	Medium Grid	Coarse Grid
Total number of cells	7,254,528	906,816	113,352
Surface cells in chordwise direction	96	48	24
Surface cells in spanwise direction	96	48	24

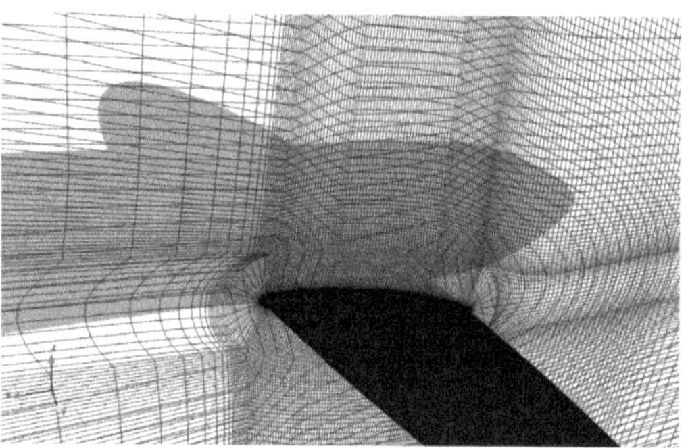

Figure 4. A detail of the grid near the foil.

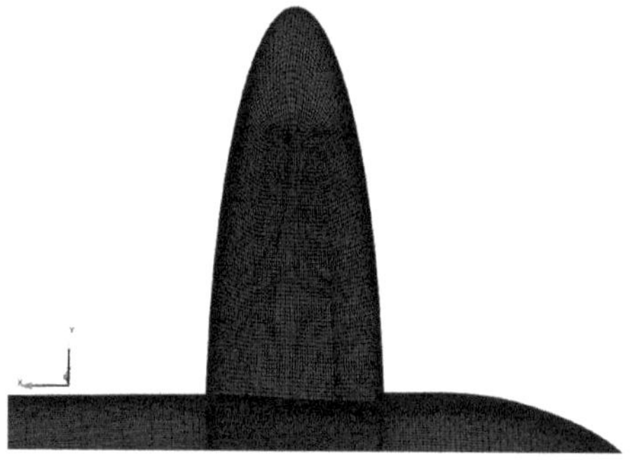

Figure 5. Grid resolution on the surface of the foil. Fine grid.

The foils and the body are modelled as no-slip solid surfaces. A velocity boundary condition is applied at the inlet, and a pressure boundary condition is applied at the outlet. Slip boundary condition is applied at the tunnel walls. The inflow velocity is set based on the case considered, and the background pressure level is set to either a large value for the wetted flow simulations, $\mathcal{O}(10^6)$, or set based on the cavitation numbers in the cavitating cases. The calculations can be typically performed on three grid levels. On the coarser grid levels, every second point is taken into account compared to the finer level grid. The computations on a coarser grid are used as the initial guess for the computations performed on a finer grid. Initially, we made the unsteady computations for the coarse and medium size meshes to keep the computational time within reasonable limits. However, for the $\sigma = 1.27$ case, the cavitation extent is not as wide as in the $\sigma = 0.53$ case, and the dynamics appear at a higher frequency. To capture better cavitation dynamics, we considered it adequate to also use a finer mesh in the simulations.

4. Results

4.1. Cavitation Observation

Cavitation observations made in the experiments in the case with $U_\infty = 8$ m/s and $\sigma = 0.53$ are shown in Figure 6, and a snapshot of the simulations with the SST model is given in Figure 7, where the pressure coefficient is defined as $C_p = 2p_{dif}/\rho_\infty U_\infty^2$, where p_{dif} denotes the pressure difference. Generally, attached sheet cavitation covers roughly half of the suction side in the experiments with cavitation incepting at the leading edge. Almost the entire suction side is covered by unstable detached sheet cavitation in the simulations. Strong and periodic shedding of the sheet cavitation are observed in the simulations, with cloud-like cavities shedding to the wake of the foils. In the experiments, the sheet cavitation is observed to break off as cloud cavitation. We observed streak cavitation at the hemispherical head of the cavitator body in the experiments.

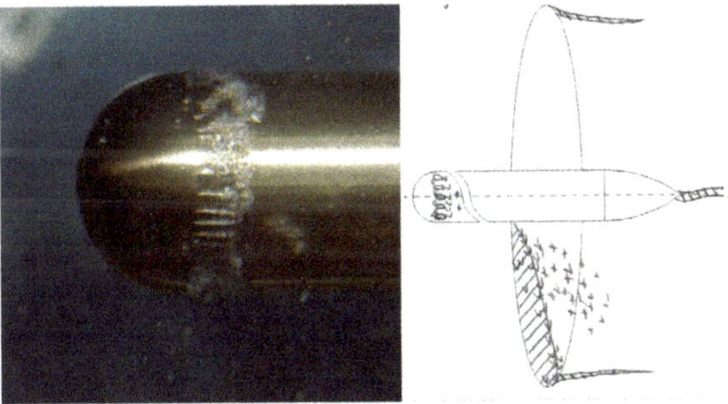

Figure 6. Cavitation extent observed in the experiments with $U_\infty = 8$ m/s and $\sigma = 0.53$. Left picture shows a photograph of the head of the cavitator in the experiments, and right picture shows a sketch of the mean cavitation extent on the cavitator. The pictures are reproduced with permission from Rhena Klose (SVA).

In the simulations, however, we see attached sheet cavitation at the hemispherical head of the cavitator body. Strong hub vortex cavitation extending far behind the cavitator body is also observed both in the experiments as well as in the simulations. We notice that in the experiments, a fine tip vortex cavitation extends roughly a chord length in the slipstream, while in the simulations tip vortex cavitation is not visible. In the simulations, strong disturbances created by the unsteady suction side sheet cavities are observed both in the pressure coefficient as well as in the non-dimensional v-velocity. In particular, a collapse event of a cavity cloud gave rise to a considerable pressure peak and a subsequent shock-wave-like front in the wake of the foil. The disturbances in C_p and v-velocity, however, dissipate quickly near the end of the cavitator body as the grid resolution deteriorates.

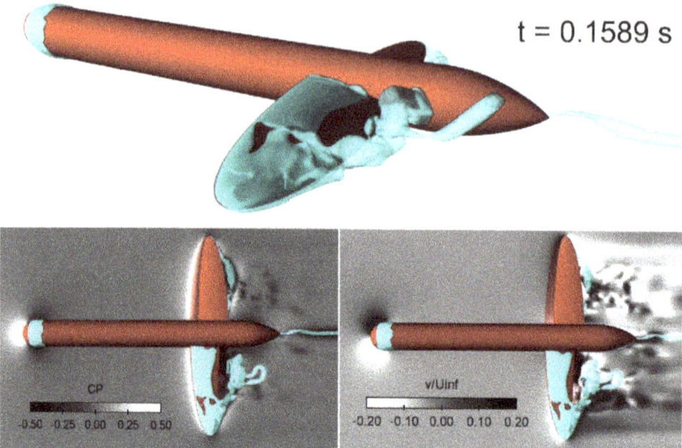

Figure 7. A snapshot of the simulations with $U_{\infty} = 8$ m/s and $\sigma = 0.53$. Fine grid with the SST turbulence model. Top frame presents an overview of the cavity extent with the transparent iso-surface of $\alpha = 0.1$ colored by blue. Lower left frame shows the pressure coefficient on the center z plane of the cavitator. Lower right frame shows the non-dimensional v-velocity on the center z plane of the cavitator.

Cavitation observations made in the experiments in the case with $U_{\infty} = 8$ m/s and $\sigma = 1.27$ are shown in Figure 8, and a snapshot of the DDES is shown in Figure 9. Generally, sheet cavitation covers roughly half of the suction side in the experiments, and roughly one third of the suction side in the simulations with cavitation incepting at the leading edge. Periodic shedding of the sheet cavitation into smaller cavity structures is observed in the simulations. These smaller cavities convected a distance along the foil before disappearing. In the experiments, parts of the sheet cavitation are observed to break off as cloud cavitation. A cavitating hub vortex is visible both in the experiments as well as in the simulations. We noticed that in the experiments, a very fine tip vortex cavitation extends roughly a chord length in the slipstream, while in the simulations tip vortex cavitation is not visible. In the simulations, disturbances created by the unsteady suction side sheet cavities are observed both in the pressure coefficient as well as in the non-dimensional v-velocity. A collapse event of the smaller cavity structures gave rise to a pressure peak and a subsequent shock-wave-like front in the wake of the foil. The disturbances, however, dissipate quickly near the end of the cavitator body as the grid resolution deteriorates.

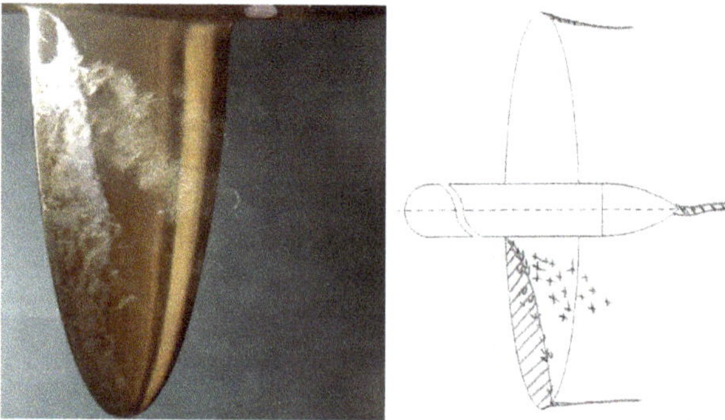

Figure 8. Cavitation extent observed in the experiments with $U_{\infty} = 8$ m/s and $\sigma = 1.27$. Left picture shows a photograph of the right foil of the cavitator in the experiments, and right picture shows a sketch of the mean cavitation extent on the cavitator. The pictures are reproduced with permission from Rhena Klose (SVA).

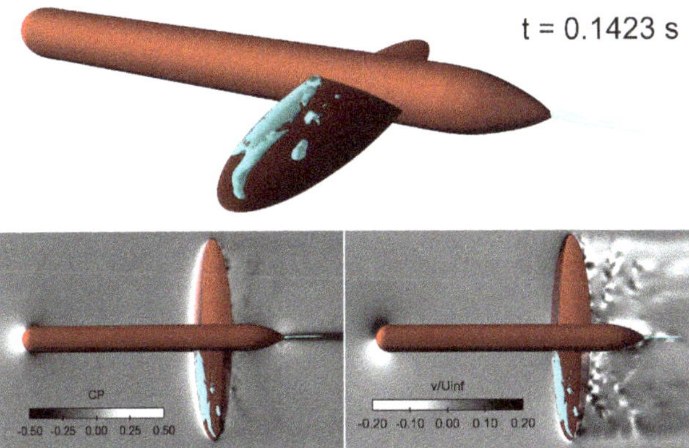

Figure 9. A snapshot of the simulations with $U_\infty = 8$ m/s and $\sigma = 1.27$. Fine grid DDES. Top frame presents an overview of the cavity extent with the transparent iso-surface of $\alpha = 0.1$ colored by blue. Lower left frame shows the pressure coefficient on the center z plane of the cavitator. Lower right frame shows the non-dimensional v-velocity on the center z plane of the cavitator.

4.2. Global Forces

The mean values of the lift and drag coefficients were extracted from the simulations. In addition, we carried out a Fourier analysis for the lift and drag forces on the wings to determine the unsteady characteristics of especially the sheet-cavitation shedding. The results are shown in Table 2 for all investigated cases.

Table 2. Comparison of the mean lift and drag coefficients ($C_{\mathscr{L}}$ and $C_{\mathscr{D}}$, respectively), and their dominant frequencies (f_0).

Case	$\overline{C}_{\mathscr{L}}$	$\overline{C}_{\mathscr{D}}$	Lift: f_0 (Hz)	Drag: f_0 (Hz)
$U_\infty = 8.0$ m/s, $\sigma = 0.53$				
Left foil: coarse grid, SST	0.078	0.025	29.0	29.0
Right foil: coarse grid, SST	0.078	0.025	29.0	29.0
Left foil: coarse grid, DDES	0.079	0.025	29.0	29.0
Right foil: coarse grid, DDES	0.079	0.025	29.0	29.0
Left foil: medium grid, SST	0.071	0.023	29.0	29.0
Right foil: medium grid, SST	0.071	0.023	29.0	29.0
Left foil: medium grid, DDES	0.075	0.024	33.4	35.0
Right foil: medium grid, DDES	0.075	0.024	33.4	35.0
$U_\infty = 8.0$ m/s, $\sigma = 1.27$				
Left foil: coarse grid, DDES	0.110	0.019	-	-
Right foil: coarse grid, DDES	0.109	0.019	-	-
Left foil: medium grid, DDES	0.121	0.017	160.2	160.2
Right foil: medium grid, DDES	0.120	0.017	160.2	160.2
Left foil: fine grid, DDES	0.118	0.016	159.6	159.9
Right foil: fine grid, DDES	0.117	0.016	159.8	160.2

We observed a dominant frequency of around 30 Hz for the case with and $\sigma = 0.53$. We show an example of the time histories and corresponding frequency contents in Figure 10 obtained from DDES, and in Figure 11 obtained from the SST simulations. A frequency of 29.0 Hz for both the lift and drag coefficients was seen with coarse and medium grids when using the SST model. A dominant frequency of 29.0 Hz was observed with coarse grid DDES for both the lift and drag coefficients, while a dominant frequency of 33.4 Hz was observed for lift and 35.0 Hz was observed for the drag with medium grid DDES. Animations from fine and medium grid DDES and SST simulations revealed a frequency of approximately 30 Hz for the sheet-cavitation shedding. Comparing the DDES and SST simulations, we see that the shedding frequency increases slightly with DDES on the medium grid. The time histories of the SST simulations appear slightly more regular, which is seen also as the sharper frequency peaks whereas these are more spread when using DDES. If we approximate the reference length of the Strouhal number as the chord length, the Strouhal numbers' range is $St = fc/U_{\infty} = [0.22 \ldots 0.26]$. If we instead take the reference velocity as $U_{ref} = U_{\infty}\sqrt{1+\sigma}$, the Strouhal numbers would range as $St = [0.18 \ldots 0.21]$.

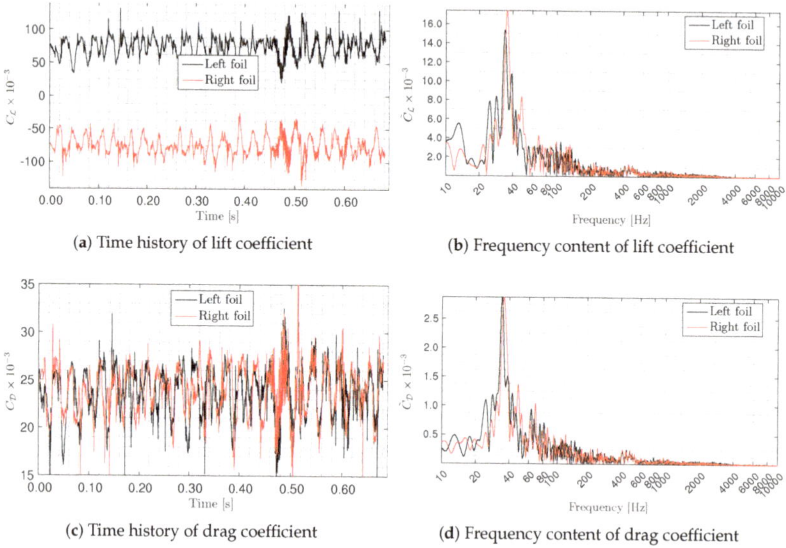

(a) Time history of lift coefficient (b) Frequency content of lift coefficient

(c) Time history of drag coefficient (d) Frequency content of drag coefficient

Figure 10. Time history and frequency content of the lift and drag coefficients. Medium grid, $\sigma = 0.53$, DDES.

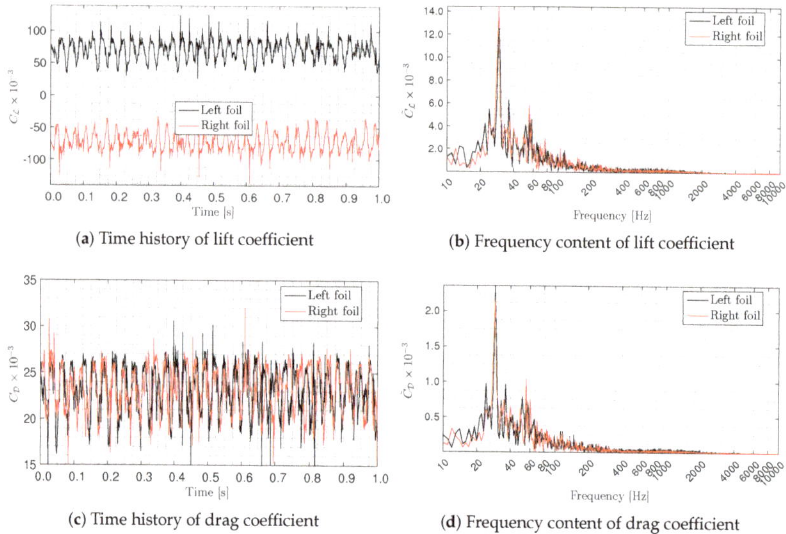

(a) Time history of lift coefficient (b) Frequency content of lift coefficient

(c) Time history of drag coefficient (d) Frequency content of drag coefficient

Figure 11. Time history and frequency content of the lift and drag coefficients. Medium grid, $\sigma = 0.53$, SST simulations.

For the case with $\sigma = 1.27$ a dominant frequency in between 159.6–160.2 Hz for both the lift and drag coefficients was observed with medium and fine grid DDES. As we discuss later in Section 4.6, the SST model yielded a steady-state solution for this case. The time histories and corresponding frequency contents are shown in Figure 12. Also, the animations created from fine and medium grid DDES revealed a frequency of approximately 160 Hz for the sheet-cavitation shedding. If we approximate the reference length of the Strouhal number as a quarter of the chord length, the Strouhal number would be $St = fc/U_\infty = 0.30$. If instead we take the reference velocity as $U_{ref} = U_\infty\sqrt{1+\sigma}$, the Strouhal number would be $St = 0.20$. Compared to the case with the lower cavitation number, we now see more distinct peaks even up to the 4th harmonics of the lift and drag.

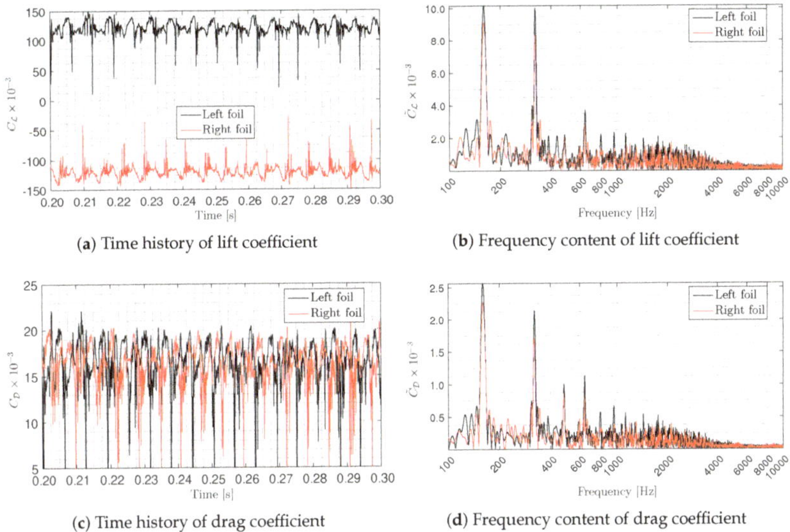

(**a**) Time history of lift coefficient
(**b**) Frequency content of lift coefficient
(**c**) Time history of drag coefficient
(**d**) Frequency content of drag coefficient

Figure 12. Time history and frequency content of the lift and drag coefficients. Fine grid, $\sigma = 1.27$, DDES.

We note that the conducted simulation time was not enough with the fine grid for the case with $\sigma = 0.53$ to provide a sufficient time history to conduct a reliable Fourier analysis. The resolution on the coarse grid was not enough to obtain a reasonable frequency for the cavitation shedding with $\sigma = 1.27$.

There was a small deviation between the mean lift coefficients of the left and right foils with the same cavitation number, while the mean drag coefficients were the same for both foils on each grid. We noted small differences also in the frequencies exhibited in the time histories. A decrease of approximately 40% of the mean lift and an increase of approximately 40% of the mean drag are seen in the results with $\sigma = 0.53$ compared to the results of $\sigma = 1.27$. This is because there is considerably more suction side sheet cavitation for the case with $\sigma = 0.53$, since the cavitation number is lower. The increase in the amount of the vapor phase on the suction side in turn decreased the lift and increased the drag of the foil. This difference was visible in the results using each of the three different grid densities.

4.3. Analyses of Cavitation Dynamics

Next, we compare the simulations to the experiments in Section 4.3.1 in terms of the hydrophone measurements. The dynamics of the investigated cases with different cavitation numbers are then analyzed in Sections 4.3.3 and 4.3.4. Cavitation collapse events and resulting pressure variations are investigated in Section 4.4. Later in Section 4.5 we demonstrate a formation of a cavitating ring vortex, and in Section 4.6 a comparison of the DDES and SST simulations is shown.

4.3.1. Comparisons with Hydrophone Measurements

Different hydrophone configurations were tested in the cavitation tunnel of SVA Potsdam, depicted in Figure 13 and summarized in Table 3. The hydrophones were installed on the starboard side of the cavitation tunnel window.

Table 3. Hydrophone arrangements in the cavitation tunnel.

Hydrophone	Installation
HF1	Upstream, acoustic chamber
HF2	Downstream, acoustic chamber
HF3	Upstream, longitudinally to the flow
HF4	Downstream, longitudinally to the flow
HF5	Upstream, transversal to the flow
HF6	Downstream, transversal to the flow

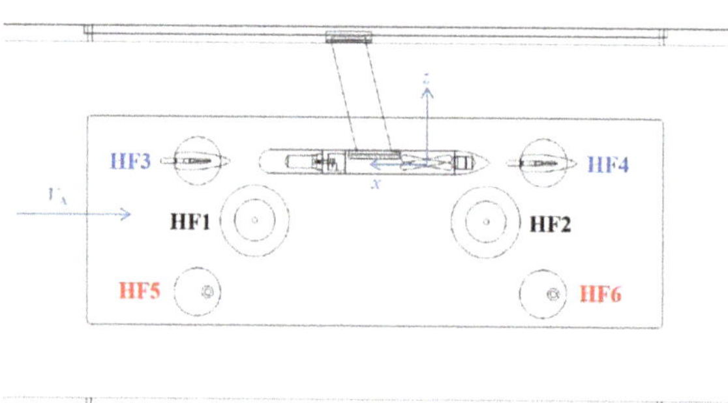

Figure 13. Hydrophone arrangement in the starboard side of the cavitation tunnel. Figure adapted from [32], and reproduced with permission from Rhena Klose (SVA).

Figure 14 shows the results of the conducted hydrophone measurements. In the figures, we show six different hydrophone arrangements. HF1 and HF2 denote the hydrophones in an acoustic chamber, with HF1 mounted in the wall of the cavitation tunnel at a distance of 0.500 m from the origin of the cavitator (see Figure 13), and HF2 mounted a distance of 0.414 m from the cavitator. HF3 and HF4 are located longitudinally to the flow, and the distances to the cavitator are 0.543 m for HF3 m and 0.352 m for HF4. HF5 and HF6 are located transversally to the flow, and the distances to the cavitator are 0.522 m for HF5 m and 0.420 m for HF6. The sensors HF1, HF3 and HF5 are located upstream of the cavitator, and the sensors HF2, HF4 and HF6 are located downstream of the cavitator.

Figure 14a shows the case with $\sigma = 0.53$, and Figure 14b shows the case with $\sigma = 1.27$. In both instances, the predicted sheet-cavitation shedding frequencies and their three harmonics are shown as well, cf. Section 4.2. We observe that in the case of $\sigma = 0.53$, the shedding frequency is picked up by hydrophones HF1, HF2, HF3 and HF4. Moreover, the measurements conducted with HF2, HF3 and HF4 show a smaller peak at the second harmonic of the sheet-cavitation shedding frequency. In the case of $\sigma = 1.27$, we see that the CFD predicted cavitation shedding frequency peak is visible only in the HF1 measurements. Other frequencies are also seen in the hydrophone measurements that are of greater magnitude than the ones induced by cavitation dynamics in the CFD simulations.

For instance, the shedding cavities formed into cloud-like cavitation structures both in the experiments as well as in the simulations, in both the investigated operation points. This should manifest itself as a high-frequency source of noise, the type of which is seen in the hydrophone signals in Figure 14: at frequencies between approximately 100 Hz and 1 kHz for the case with $\sigma = 0.53$ and at frequencies around 1–2 kHz for $\sigma = 1.27$, visible especially in the hydrophones HF1 and HF2. The high-frequency broadband noise is likely due in part to very rapid collapse events of the cavity clouds once they reach a region of sufficiently high pressure in the slipstream of the cavitator. A cavitation collapse event is assessed in Section 4.4. The best arrangement of hydrophones in the tests for the investigations with the standard cavitator turned out to be the one in the acoustic chamber, i.e., hydrophones HF1 and HF2.

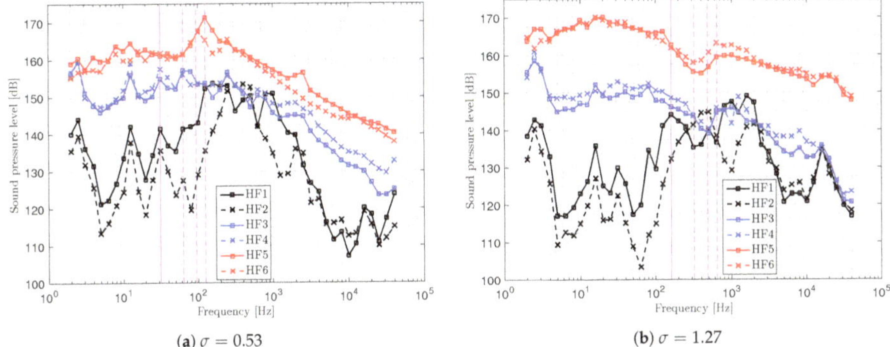

(a) $\sigma = 0.53$ **(b)** $\sigma = 1.27$

Figure 14. Sound pressure level measurements inside the cavitation tunnel with different hydrophone arrangements. The magenta vertical lines denote the predicted cavitation shedding frequency and its harmonics. The measurement results were reproduced with permission from Rhena Klose (SVA).

4.3.2. Cavitation Dynamics and Shedding Mechanisms

Below in Sections 4.3.3 and 4.3.4, we discuss observations of the sheet-cavitation dynamics and shedding mechanisms. Figure 15 shows a sheet-cavitation shedding cycle of the case with $\sigma = 0.53$. Figure 16 shows a sheet-cavitation shedding cycle of the case with $\sigma = 1.27$. Former The results are obtained for the former on the medium grid, and the latter on the fine grid. We study here the cavitation dynamics by visualizing on the left side of a frame the rate of evaporation together with velocity vectors on a cut plane at mid-span, as well as the contour of void fraction value $\alpha = 0.1$. A negative rate of evaporation denotes the rate of condensation. Additionally, on the right side of the frame a perspective view of the cavitation extent is shown by an iso-surface of $\alpha = 0.1$. In each figure, six frames are presented to represent one cavitation shedding cycle. The shedding cycle is taken as the corresponding frequency shown earlier in Table 2. It is notable that the flow features and cavitation dynamics, especially in the case with the lower cavitation number, were observed as highly three-dimensional over the span of the foil. Nevertheless, we focus on a qualitative sense on the flow features of the aforementioned cut plane at mid-span.

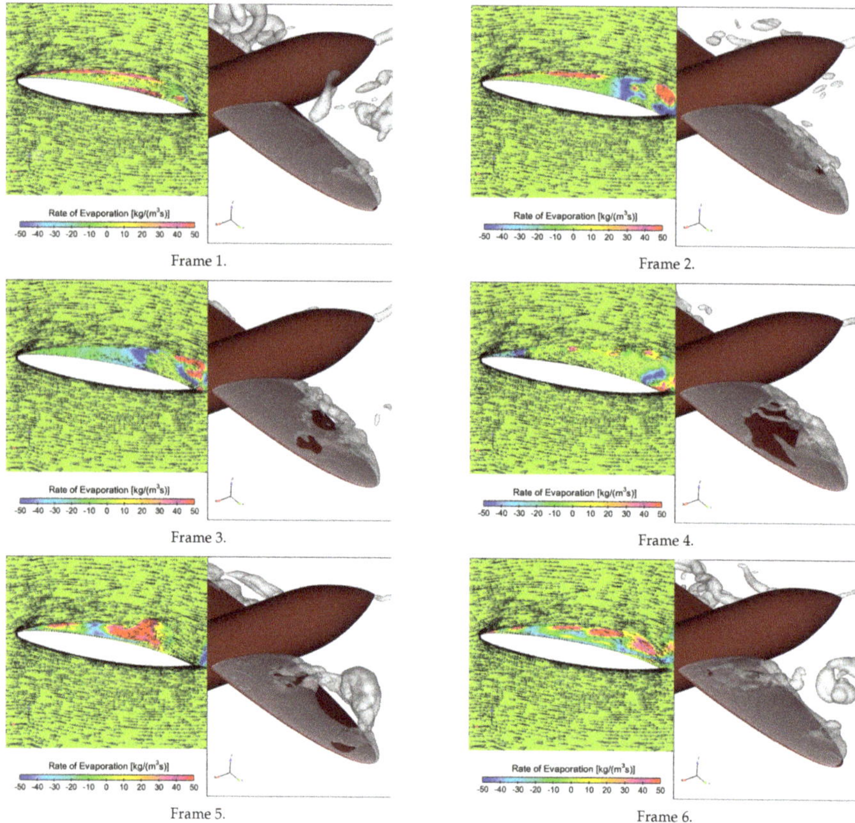

Figure 15. A sequence of snapshots visualizing one suction side sheet-cavitation shedding cycle. DDES on medium grid, $\sigma = 0.53$. Left frame depicts a cut plane of the left foil at mid-span, with velocity vectors denoted by arrows (only 25% of the velocity vectors is drawn on the plane for clarity), contour of the void fraction $\alpha = 0.1$ denoted by the grey curves, and the plane is colored by the rate of evaporation. The right frame shows the cavity extent near the left foil by an iso-surface of $\alpha = 0.1$. The frame number denotes the respective stage of the cycle.

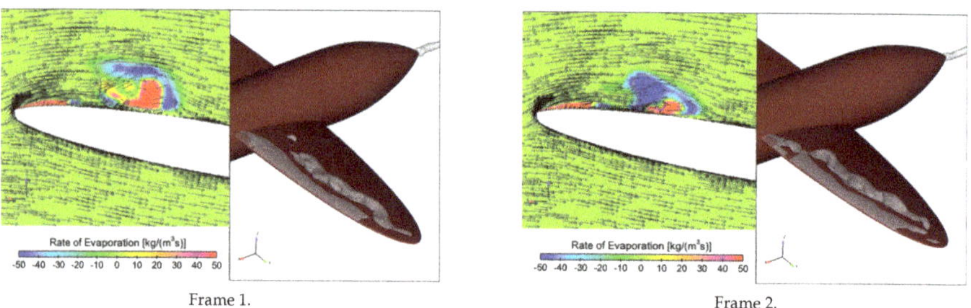

Figure 16. *Cont.*

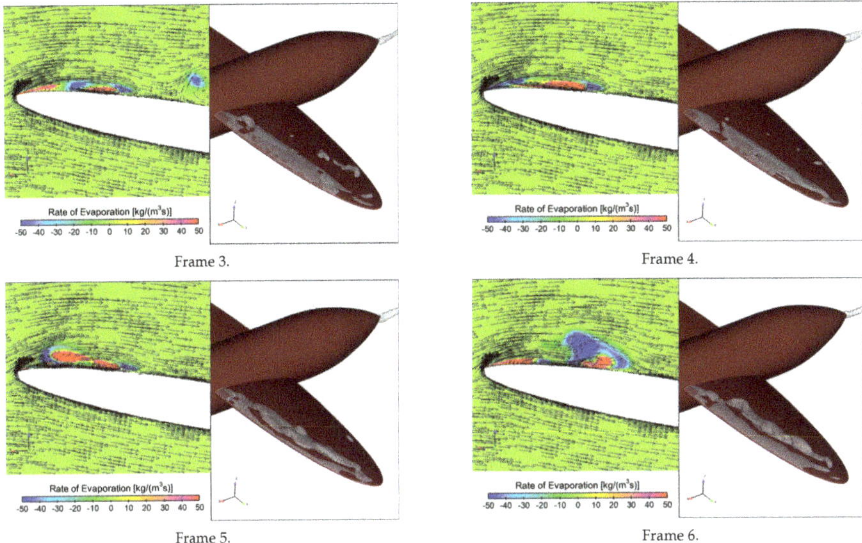

Figure 16. A sequence of snapshots visualizing one suction side sheet-cavitation shedding cycle. Fine grid, $\sigma = 1.27$, DDES. Left frame depicts a cut plane of the left foil at mid-span, with velocity vectors denoted by arrows (only 25% of the velocity vectors is drawn on the plane for clarity), contour of the void fraction $\alpha = 0.1$ denoted by the grey curves, and the plane is colored by the rate of evaporation. The right frame shows the cavity extent near the left foil by an iso-surface of $\alpha = 0.1$. The frame number denotes the respective stage of the cycle.

4.3.3. Cavitation Dynamics: $\sigma = 0.53$

In the first frame of Figure 15 (case $\sigma = 0.53$) a cavity cloud is shed in the wake of the foil, and an attached sheet cavity has grown from the leading edge of the foil. There is strong evaporation in the layers close to the wall of the sheet cavity and on the upper surface of the cavity. Some condensation takes place at the closure of the sheet cavity. Parts of the shed cavity clouds convect still farther downstream, visible in the right side of the first frame. The frames 1–2 depict the growth of the attached sheet cavitation to super-cavitation, where the sheet cavity momentarily covers the entire suction side of the foil. A re-entrant jet is observed at the closure line of the sheet cavitation in frame 2. Once we reach Frame 3, the re-entrant jet has impinged the sheet cavitation, and continues to a draw liquid phase underneath it. We also notice strengthening condensation at the closure of the sheet cavitation in frames 2–3, as well as strong vorticity as the shedding initiates. By Frame 4, almost all sheet cavity has detached at mid-span, and the re-entrant jet has further intensified. We observe that the impingement does not occur simultaneously over the whole span, but rather exhibits a highly three-dimensional shape and structure with the root and tip parts of the foil still completely covered by the sheet cavitation. As we reach Frame 5, a large cavity cloud is broken off the sheet cavity, which is convected to the wake of the foil. Strong evaporation begins again on the foil as we notice at the cut plane. Finally, once we reach Frame 6 the sheet cavitation at the LE continues to grow, and the broken-off cavity clouds are convected farther in the wake by the external flow over the foil. A new shedding cycle is then initiated at Frame 6. Again we note that while the visually observed shedding cycle was confirmed by the dominant frequencies of the lift and drag coefficients, the two-phase flow over the span is three-dimensional and, for instance, separate and phase-lagging breaking cycles were observed near the root parts of the foil.

4.3.4. Cavitation Dynamics: $\sigma = 1.27$

Similarly as above, the first frame of Figure 16 (case $\sigma = 1.27$) depicts the initial stages the attached sheet cavity growth. Cavity clouds are shed in the wake of the foil, and an attached sheet cavity begins to grow from the leading edge of the foil. There is strong evaporation in the layers close to the wall of the sheet cavity, and condensation takes place on the closure of the forming cavity. A detached cavity cloud is convected to the slipstream and vortical flow is associated with it. In frames 2–3, the attached sheet cavitation grows in length along the foil. In Frame 2, we observe evaporation at the center of the attached cavity cloud, whereas condensation takes place around the closure of the sheet cavity. As we reach Frame 3 a re-entrant jet is visible. The jet draws the liquid phase toward the LE, underneath the sheet cavity and splits the evaporation zone of the previous frame in two parts. Frame 4 depicts a partial break-off stage as the re-entrant jet begins to impinge the sheet cavity. Condensation now takes place both in the LE and the TE of the detaching cavity, although evaporation is present at its core. A tiny cavity at the LE of the foil begins to form. Once we reach Frame 5, the impingement of the re-entrant jet is more complete, and we observe that the sheet cavity has broken off on a wide span. The cavity cloud starts to rise away from the foil to be convected by the external flow. Stronger vortical flow forms around the core of the detached cavity. In Frame 6 we observe this cavity cloud being convected farther in the wake, as a new shedding cycle is about to begin. We furthermore observe condensation beginning to dominate over evaporation in the detached cavity. A strong recirculating flow is visible around the detached cavity cloud. Again, the flow is three-dimensional over the span of the foil, and the break-off does not occur simultaneously over the span.

4.4. *Cavitation Collapse Events and Pressure Waves*

In Figure 17 we show a set of snapshots for a stage of the shedding cycle at which the cavity undergoes a rapid collapsing stage, and the associated instantaneous pressure distributions. The figure shows the case with the lower cavitation number, $\sigma = 0.53$. The cavity cloud has shed off at t_0, and we observe an increase in pressure on the plane at mid-span, just below the cavity. As the collapse stage advances, the pressure values grow larger, and the pressure front starts to propagate in all directions quite rapidly. We note very high values for the C_p due to the major collapse events $t_0 + 0.8$ ms. The maximum value recorded for the investigated plane was ≈ 3. After this, the pressure wave front grows and intensifies, and then also quickly dissipates. Next, we are left with small regions of negative pressure in the place of the shed cavity clouds at the two latest investigated time instants. Looking at the cavitator body, the pressure wave due to the major collapse at $t_0 + 0.8$ ms reaches it a few tenths of ms later. The cavitator body on the upstream side of the wing shows areas of increased positive pressure, and the downstream side of it in turn has a momentarily lower negative pressure. At the moment of the collapse event, also high pressures are recorded on the suction side of the wing near the trailing edge, as the cavity collapses just above it, at the same time the pressure side of the wing is masked from this effect. The dynamics of the bulk cavitation on the wing, then again, appears to be unaffected by the pressure waves that result from the cavity collapse. The attached sheet cavitation continues to grow on to enter the shedding phase described in Section 4.3.3. Also, at subsequent time instants, the shed-off cavity continues to form the cavitating ring vortex. We assess this further in Section 4.5. The cavitating vortex due to the hub of the cavitator body also appears to be not affected by the collapse event.

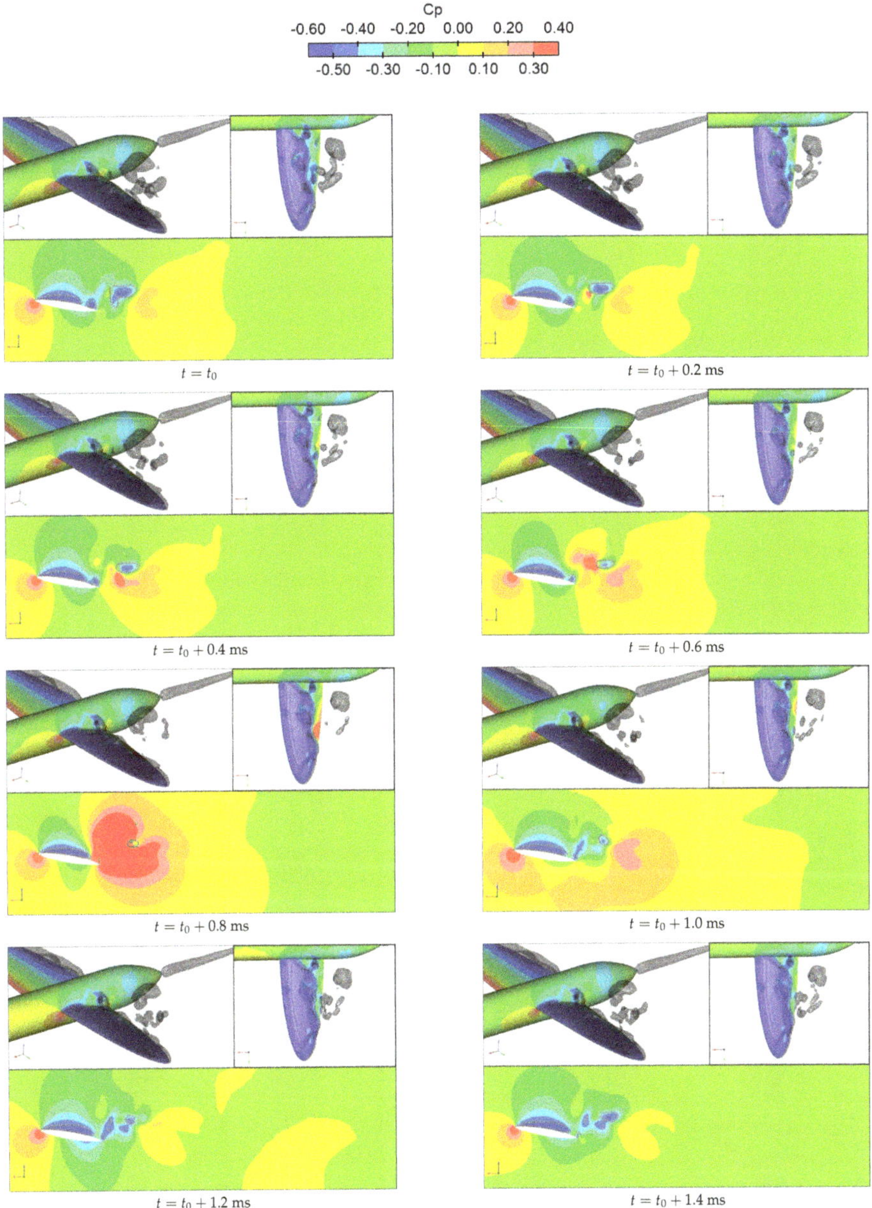

Figure 17. A sequence of snapshots visualizing a collapse event of the cavity. DDES with medium grid, $\sigma = 0.53$. Top frames show the left wing from the side and the top, and the bottom frame shows a cut plane at mid-span. The cavitator and the cut plane are colored by C_p.

4.5. A Formation of Cavitating Ring Vortex

In Figure 18 we demonstrate the formation of a cavitating ring vortex. The figure depicts six snapshots at 2 ms intervals of the DDES on a medium grid with $U_\infty = 8\ \mathrm{m/s}$ and $\sigma = 0.53$. On the left side of each frame, the isosurfaces of $\alpha = 0.1$ and Q = 40,000 are shown. Moreover, the Q-criterion is colored by helicity $\mathrm{H} = u_i\Omega_i/(|u_i||\Omega_i|)$. The helicity denotes the cosine of the angle between the velocity and the vorticity vectors, and tends to ±1 in the cores of free vortices, the sign indicating the direction of swirl. On the right side of each frame, three cut planes are shown at spans of $y/s = [0.25, 0.50, 0.75]$, which are colored by the rate of evaporation R. Also, on the right side of each frame, the iso-surface of $\alpha = 0.1$ is shown as the transparent orange surface. At the beginning of this series of snapshots ($t = t_0$), we are approaching the end of a shedding cycle (cf. Figure 15) with cavitation and vortical flow structures being shed in the wake of the foil. A complex interaction with the shedding vapor clouds and liquid-phase flow introduce vast vortical flow structures when the sheet cavity detachment takes place between $t = t_0$ and 2 ms. As separate and partially cavitating vortex filaments convect downstream, some of these join together to form a ring-shaped structure, or a toroidal-type vortex between $t = t_0 + 2$ and $\ldots 6$ ms. Observing the helicity of these toroids, we note that once it is a true vortex ring or even a piece of such, the vorticity is strong enough for the flow inside these rings to cavitate in most parts clearer and at a higher intensity. Once these structures convect farther downstream ($t = t_0 + 8 \ldots 10$ ms), dissipation owing to the coarsening grid in the wake diminishes the strength of the toroid-like shape, leading to dissipation of the cavitation as well, despite the used compressive flux limiter. Moreover, as was mentioned above, we observe the strongly three-dimensional cavitation structure in Figure 18. For instance, the rates of evaporation on the three cut planes are vastly different from each other, albeit the sheet cavitation covering the entire foil superficially appears as rather steady at several time instants. Also chord- and spanwise fluctuations can be observed on the iso-surface of the Q-criterion.

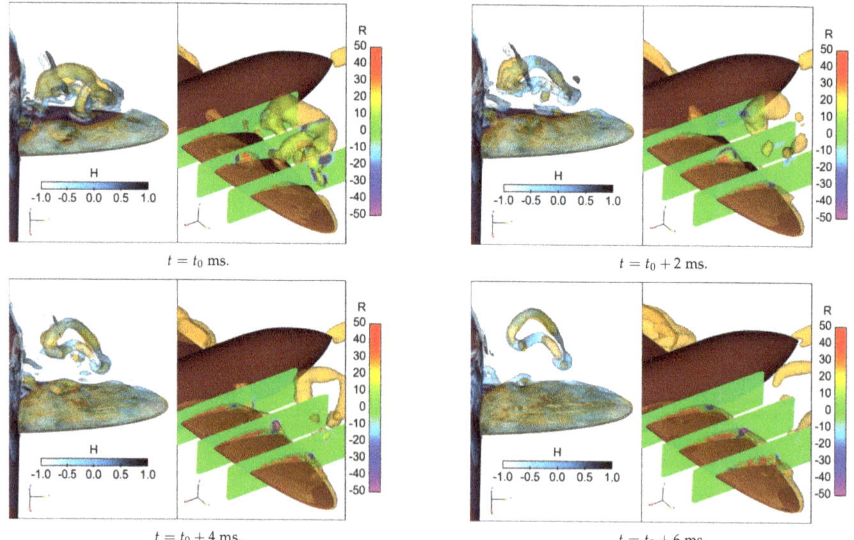

$t = t_0$ ms.

$t = t_0 + 2$ ms.

$t = t_0 + 4$ ms.

$t = t_0 + 6$ ms.

Figure 18. *Cont.*

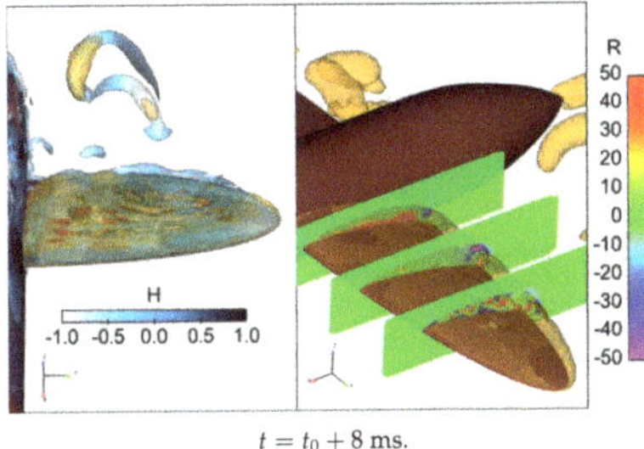

$t = t_0 + 8$ ms.

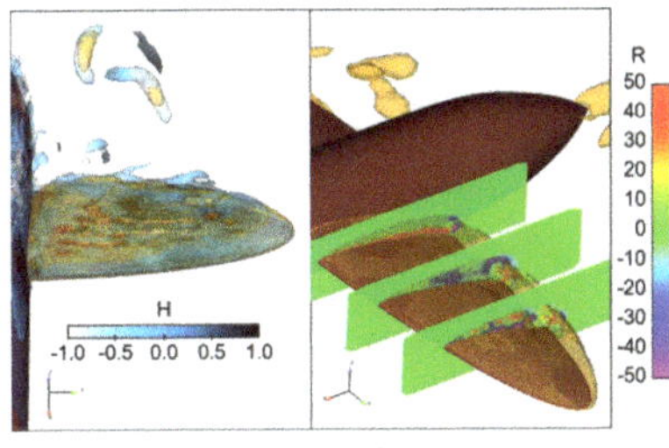

$t = t_0 + 10$ ms.

Figure 18. A sequence of snapshots depicting the formation of a partial cavitating ring vortex in the wake of the foil. DDES with medium grid, $\sigma = 0.53$. On the left in each frame: left foil with iso-surface of void fraction $\alpha = 0.1$ colored as orange, together with an iso-surface of $Q = 40,000$, which is colored by the helicity H. On the right in each frame: three cut planes at $y/s = [0.25, 0.50, 0.75]$ which are colored by the rate of evaporation R, with iso-surface of void fraction $\alpha = 0.1$ colored as orange.

4.6. A Comparison of Different Turbulence Closures

A peculiar modelling issue regarding the cavitation dynamics was observed during the simulations. The case with $\sigma = 1.27$ was initially begun with the SST model, but the predicted suction side sheet cavitation remained stable throughout the simulations. We observed no natural shedding of the sheet cavitation, and the time histories of the global forces were flat. We carried out the simulations with the SST model until $t \approx 0.155$ s, after which the DDES approach was activated. The outcome is illustrated in Figure 19 in terms of the time histories of the lift and drag coefficients. It was observed that almost immediately after the DDES approach was activated, the suction side sheet cavitation began its shedding behavior, leading to high-frequency oscillations also in the global force coefficients.

A comparison of the DDES and the SST simulation is shown in Figure 20, made from visualizations of the eddy-viscosity levels near the LE of the foil on a cut plane at mid-span, cavitation extents and vortical flow structures. We observe a significant difference in the eddy-viscosity distributions. The SST simulation predicts several orders of magnitude larger eddy-viscosity levels on the suction side of the foil. This not only suppresses any vortical structures around the foil, but also prevents the natural shedding mechanisms of the sheet cavitation to take place. Instead, the predicted suction side cavitation remained stable as consequently did the lift and drag forces as well, as we noted in Figure 19. By contrast, when using DDES, similar levels of eddy-viscosity are not predicted by the model at the closure of the sheet cavity. We also note considerably more unsteady vortical flow structures with DDES, shed in the wake of the foil. Due to these reasons, the natural shedding of the sheet cavity did occur as was demonstrated in Sections 4.2 and 4.3.

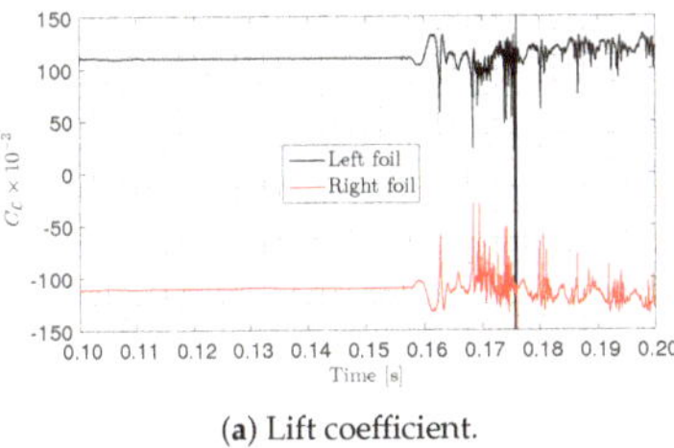

(**a**) Lift coefficient.

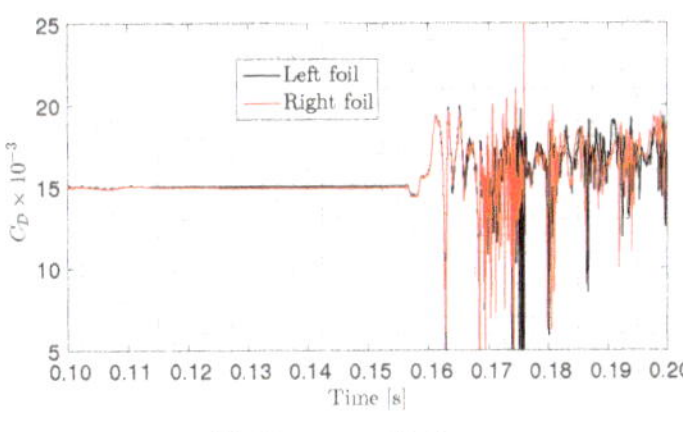

(**b**) Drag coefficient.

Figure 19. Illustration of an effect of DDES. Simulations were carried out using SST $k - \omega$ until $t \approx 0.155$ s, after which the DDES approach was activated. Medium grid, $U_\infty = 8$ m/s, $\sigma = 1.27$.

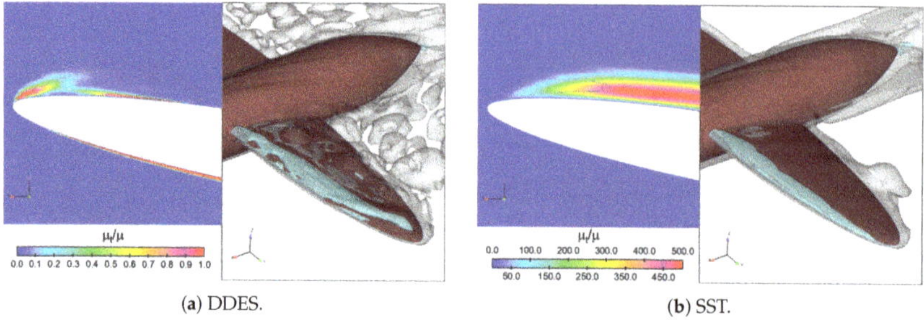

(a) DDES. (b) SST.

Figure 20. Snapshots of simulations with DDES and SST. Medium grid, $\sigma = 1.27$. Left frame depicts a cut plane of the left foil at mid-span with the contour of void fraction $\alpha = 0.1$ denoted by the grey curves, and the plane is colored by the eddy-viscosity ratio. The right frame shows the cavity extent near the left foil by an iso-surface of the void fraction $\alpha = 0.1$, colored by light blue, and vortical flow structures by an iso-surface of $|\Omega_i| = 300$ 1/s as the transparent grey color.

We note here the natural shedding behavior of the case with lower cavitation number was reasonably captured also with the used URANS approach, as discussed in Section 4.2 and seen in Figures 10 and 11. This observation might be because the foil was, in an average sense, at a super-cavitating state where the attached sheet cavitation momentarily covered the entire suction side of the foil. It appears that in this case the shedding mechanism is strong enough to overcome the, perhaps excess, diffusion originating from the eddy-viscosity modelling even on the medium grid.

5. Conclusions

We studied the ITTC Standard Cavitator numerically in a cavitation tunnel at different cavitation numbers. To our knowledge, our numerical analyses are the first reported ones for the ITTC standard cavitator. Emphasis was put upon the assessment of sheet-cavitation dynamics. A compressible two-phase flow model was used for the flow solution, and two turbulence closures are employed: a two-equation RANS model, SST $k - \omega$, and a hybrid RANS/LES model, called the delayed-detached eddy simulation (DDES). A homogeneous mixture model was used for the two phases.

We compared the simulations with observations made in the cavitation tunnel experiments. Good agreement was achieved with the mean cavitation extent, in terms of the sheet cavitation on the foils, and the hub vortex cavitation which formed behind the cavitator body. The numerically predicted sheet-cavitation shedding frequencies can also be observed with some hydrophones in the acoustic measurements, which were carried out in the cavitation tunnel. Streak cavitation at the hemispherical head of the cavitator body was predicted as an attached sheet cavitation in the simulations. The numerical grid resolution did not permit the capturing of the very fine tip vortex cavitation that was observed in the experiments. Also, the prediction of flows with bubble cavitation is difficult with a homogeneous two-phase mixture model.

Detailed analysis of the cavitation shedding mechanism confirmed that the dynamics of the sheet cavitation are dictated by the re-entrant jet. Once the jet impinges the sheet cavitation, a detachment takes place together with complex interaction between the re-entrant jet, vortical flow structures and mass transfer near the phase boundary. We observed that the break-off cycle was relatively periodic in both investigated cases with an approximately constant shedding frequency. The Strouhal numbers lay within the usual ranges reported in the literature for sheet-cavitation shedding. Furthermore, we noted that a strong vortical flow formed around the detached cavity. In the case of the lower cavitation number, we demonstrated that the vortical flow structures convected to the slipstream of the foils, where they developed striking cavitating toroidal vortices. An intricate interplay with the shed vapor clouds and the liquid-phase flow introduced vast vortical flow structures when the sheet cavity detachment took place. We showed that once the toroidal vortex ring was formed, the vorticity was in most parts strong enough for the flow inside the vortex rings to cavitate. Furthermore, a collapse event of a cavity cloud in the slipstream of the foils gave rise to a considerable pressure peak and subsequent wave-like fronts.

DDES and simulations with the URANS model revealed a difference of approximately 5 Hz in the sheet-cavitation shedding for the case with the lower cavitation number. The shedding behavior was reasonably captured also with a two-equation turbulence model. This observation might be because the foil was, in an average sense, at a super-cavitating state where the attached sheet cavitation momentarily covered the entire suction side of the foil. It appears that in this case the shedding mechanism is strong enough to overcome the, perhaps excess, diffusion originating from the eddy-viscosity modelling even on the rather coarse grids. For the case with the higher cavitation number, the two-equation turbulence model predicted a steady attached sheet cavitation, whereas the hybrid RANS/LES model predicted a highly transient sheet-cavitation shedding being on a par with the experiments. We attributed this difference to the eddy-viscosity modelling. In this case, detailed vortical structures around the foil and the shedding mechanisms of the sheet cavitation were inhibited due to the unsteady RANS modelling approach.

Recently we have investigated the Reynolds number sensitiveness of different cavitation types observed on a ship's propeller [31]. For the case of static hydrofoils exhibiting similar phenomena as shown here, this remains to be assessed. Moreover, the ability of an asymptotic mass-transfer model coupled to the compressible flow solver to predict the shock wave dynamics, which can have an effect on the upstream cavitation dynamics [1], remains an open issue. One alternative is to include the Rayleigh–Plesset equation, which takes the bubble dynamics better into account, to the flow model for a more complete physical description of the two-phase flow. Other options to achieve better results comprise more holistic modelling approaches that can better account for the phase change physics, such as inhomogeneous or multi-scale flow models. An extended model of the former type is under development by the authors, and we have obtained promising initial results of propeller cavitation using such an approach.

Author Contributions: Conceptualization, V.M.V., T.S. (Timo Siikonen), A.S.-C. and T.S. (Tuomas Sipilä); methodology, V.M.V. and T.S. (Tuomas Sipilä); software, V.M.V. and T.S. (Timo Siikonen); validation, V.M.V.; formal analysis, V.M.V.; investigation, V.M.V.; resources, V.M.V. and T.S. (Tuomas Sipilä); data curation, V.M.V. and T.S. (Timo Siikonen); writing—original draft preparation, V.M.V.; writing—review and editing, T.S. (Timo Siikonen), A.S.-C. and T.S. (Tuomas Sipilä); visualization, V.M.V.; supervision, T.S. (Timo Siikonen), A.S.-C. and T.S. (Tuomas Sipilä); project administration, T.S. (Tuomas Sipilä); funding acquisition, T.S. (Tuomas Sipilä). All authors have read and agreed to the published version of the manuscript.

Funding: V.M.V. would like to express his gratitude for the support granted by the Finnish Funding Agency for Innovation (TEKES) and the German Federal Ministry for Economic Affairs and Energy within the project "PropNoise". Without the funding the research project could not have been realized. The APC was funded by VTT.

Acknowledgments: The authors would like to express their gratitude to Rhena Klose and Lars Lübke from SVA Potsdam GmbH for providing the experimental results and photographs.

Conflicts of Interest: The authors declare no conflict of interest. The funders had no role in the design of the study; in the collection, analyses, or interpretation of data; in the writing of the manuscript, or in the decision to publish the results.

Abbreviations

The following abbreviations are used in this manuscript:

CFD	Computational fluid dynamics
DES	Detached-eddy simulation
DDES	Delayed DES
ITTC	International Towing Tank Committee
LE	Leading edge
LES	Large eddy simulation
MUSCL	Monotonic upstream-centered scheme for conservation laws

RANS	Reynolds-averaged Navier–Stokes
SST	Shear stress transport
TE	Trailing edge
URANS	Unsteady RANS

References

1. Ganesh, H.; Mäkiharju, S.; Ceccio, S. Bubbly shock propagation as a mechanism for sheet-to-cloud transition of partial cavities. *J. Fluid Mech.* **2016**, *802*, 37. [CrossRef]
2. Schnerr, G.H.; Sezal, I.; Schmidt, S. Numerical investigation of three-dimensional cloud cavitation with special emphasis on collapse induced shock dynamics. *Phys. Fluids* **2008**, *20*, 040703. [CrossRef]
3. Sipilä, T. RANS Analyses of Cavitating Propeller Flows. Ph.D. Thesis, School of Engineering, Aalto University, Espoo, Finland, 2012.
4. Dular, M.; Bachert, R. The issue of Strouhal number definition in cavitating flow. *J. Mech. Eng.* **2009**, *55*, 666–674.
5. Lei, W.; Lu, C.J.; Jie, L.; Xin, C. Numerical simulations of 2D periodic unsteady cavitating flows. *J. Hydrodyn. Ser. B* **2006**, *18*, 341–344.
6. Saito, Y.; Nakamori, I.; Ikohagi, T. Numerical analysis of unsteady vaporous cavitating flow around a hydrofoil. In Proceedings of the 5th International Symposium on Cavitation (CAV2003), Osaka, Japan, 1–4 November 2003.
7. Leroux, J.B.; Coutier-Delgosha, O.; Astolfi, J. A joint experimental and numerical study of mechanisms associated to instability of partial cavitation on two-dimensional hydrofoil. *Phys. Fluids* **2005**, *17*, 052101. [CrossRef]
8. Matusiak, J. Pressure and Noise Induced by a Cavitating Marine Screw Propeller. Ph.D. Thesis, Helsinki University of Technology, Espoo, Finland, 1992.
9. Gnanaskandan, A.; Mahesh, K. Large eddy simulation of the transition from sheet to cloud cavitation over a wedge. *Int. J. Multiph. Flow* **2016**, *83*, 86–102. [CrossRef]
10. Eskilsson, C.; Bensow, R.E. Estimation of Cavitation Erosion Intensity Using CFD: Numerical Comparison of Three Different Methods. In Proceedings of the Fourth International Symposium on Marine Propulsors (smp'15), Austin, TX, USA, 31 May–4 June 2015.
11. Bensow, R.; Bark, G. Implicit LES predictions of the cavitating flow on a propeller. *J. Fluids Eng.* **2010**, *132*. [CrossRef]
12. Viitanen, V.M.; Hynninen, A.; Sipilä, T.; Siikonen, T. DDES of Wetted and Cavitating Marine Propeller for CHA Underwater Noise Assessment. *J. Mar. Sci. Eng.* **2018**, *6*, 56. [CrossRef]
13. Sakamoto, N.; Kamiirisa, H. Prediction of near field propeller cavitation noise by viscous CFD with semi-empirical approach and its validation in model and full scale. *Ocean Eng.* **2018**, *168*, 41–59. [CrossRef]
14. Ge, M.; Svennberg, U.; Bensow, R. Investigation on RANS prediction of propeller induced pressure pulses and sheet-tip cavitation interactions in behind hull condition. *Ocean Eng.* **2020**, *209*, 107503. [CrossRef]
15. Morgut, M.; Nobile, E.; Biluš, I. Comparison of mass transfer models for the numerical prediction of sheet cavitation around a hydrofoil. *Int. J. Multiph. Flow* **2011**, *37*, 620–626. [CrossRef]
16. Li, Z.; Pourquie, M.; Van Terwisga, T. A numerical study of steady and unsteady cavitation on a 2d hydrofoil. *J. Hydrodyn.* **2010**, *22*, 728–735. [CrossRef]
17. Ducoin, A.; Huang, B.; Young, Y.L. Numerical modeling of unsteady cavitating flows around a stationary hydrofoil. *Int. J. Rotating Mach.* **2012**, *2012*. [CrossRef]
18. Wang, C.; Wang, G.; Huang, B. Characteristics and dynamics of compressible cavitating flows with special emphasis on compressibility effects. *Int. J. Multiph. Flow* **2020**, *130*, 103357. [CrossRef]
19. Park, S.; Seok, W.; Park, S.; Rhee, S.; Choe, Y.; Kim, C.; Kim, J.H.; Ahn, B.K. Compressibility Effects on Cavity Dynamics behind a Two-Dimensional Wedge. *J. Mar. Sci. Eng.* **2020**, *8*, 39. [CrossRef]
20. Zhang, W.; Bai, X.; Ma, Z.; Chen, G.; Wang, Y. Compressible effect on the cavitating flow: A numeric study. *J. Hydrodyn. Ser. B* **2017**, *29*, 1089–1092. [CrossRef]
21. Budich, B.; Schmidt, S.J.; Adams, N.A. Numerical Investigation of a Cavitating Model Propeller Including Compressible Shock Wave Dynamics. In Proceedings of the Fourth International Symposium on Marine Propulsors (smp'15), Austin, TX, USA, 31 May–4 June 2015.

22. ITTC. Report of Cavitation Committee. In Proceedings of the 14th International Towing Tank Conference (ITTC), Ottawa, ON, Canada, 10–12 September 1975.
23. Miettinen, A.; Siikonen, T. Application of pressure- and density-based methods for different flow speeds. *Int. J. Numer. Methods Fluids* **2015**, *79*, 243–267. [CrossRef]
24. Viitanen, V.M.; Siikonen, T. Numerical simulation of cavitating marine propeller flows. In Proceedings of the 9th National Conference on Computational Mechanics (MekIT'17), Trondheim, Norway, 11–12 May 2017; International Center for Numerical Methods in Engineering (CIMNE): Barcelona, Spain, 2017; pp. 385–409, ISBN 978-84-947311-1-2.
25. Miettinen, A. *Simple Polynomial Fittings for Steam*; CFD/THERMO-55-2007; Report 55; Laboratory of Applied Thermodynamics, Aalto University: Espoo, Finland, 2007.
26. Spalart, P.R.; Deck, S.; Shur, M.; Squires, K.; Strelets, M.K.; Travin, A. A new version of detached-eddy simulation, resistant to ambiguous grid densities. *Theor. Comput. Fluid Dyn.* **2006**, *20*, 181–195. [CrossRef]
27. Menter, F.R. Two-equation eddy-viscosity turbulence models for engineering applications. *AIAA J.* **1994**, *32*, 1598–1605. [CrossRef]
28. Menter, F. Influence of freestream values on k-ω turbulence model predictions. *AIAA J.* **1992**, *30*, 1657–1659. [CrossRef]
29. Shur, M.; Spalart, P.; Strelets, M.; Travin, A. Detached-Eddy Simulation of an Airfoil at High Angle of Attack. In *Engineering Turbulence Modelling and Experiments 4*; Elsevier Science Ltd.: Oxford, UK, 1999; pp. 669–678.
30. Choi, Y.H.; Merkle, C.L. The application of preconditioning in viscous flows. *J. Comput. Phys.* **1993**, *105*, 207–230. [CrossRef]
31. Viitanen, V.; Siikonen, T.; Sánchez-Caja, A. Cavitation on Model-and Full-Scale Marine Propellers: Steady and Transient Viscous Flow Simulations At Different Reynolds Numbers. *J. Mar. Sci. Eng.* **2020**, *8*, 141. [CrossRef]
32. Klose, R.; Heinke, H. *Cavitation Tests and Noise Measurements with the Static Hydrofoil (ITTC Standard Cavitator)*; SVA Potsdam, Germany, Model Basin Report No. 4542; SVA (Potsdam Model Basin): Potsdam, Germany, 2017.

Article

Cavity Formation during Asymmetric Water Entry of Rigid Bodies

Riccardo Panciroli [1,*] and Giangiacomo Minak [2]

1 Engineering Faculty, Niccolò Cusano University, 00166 Rome, Italy
2 Department of Industrial Enginerrsing (DIN), Alma Mater Studiorum—Università di Bologna, Via Fontanelle, 40126 Bologna, Italy; giangiacomo.minak@unibo.it
* Correspondence: riccardo.panciroli@unicusano.it

Abstract: This work numerically evaluates the role of advancing velocity on the water entry of rigid wedges, highlighting its influence on the development of underpressure at the fluid–structure interface, which can eventually lead to fluid detachment or cavity formation, depending on the geometry. A coupled FEM–SPH numerical model is implemented within LS-DYNA, and three types of asymmetric impacts are treated: (I) symmetric wedges with horizontal velocity component, (II) asymmetric wedges with a pure vertical velocity component, and (III) asymmetric wedges with a horizontal velocity component. Particular attention is given to the evolution of the pressure at the fluid–structure interface and the onset of fluid detachment at the wedge tip and their effect on the rigid body dynamics. Results concerning the tilting moment generated during the water entry are presented, varying entry depth, asymmetry, and entry velocity. The presented results are important for the evaluation of the stability of the body during asymmetric slamming events.

Keywords: slamming; fluid-structure interaction; fluid detachment; cavitation

Citation: Panciroli, R.; Minak, G. Cavity Formation during Asymmetric Water Entry of Rigid Bodies. *Appl. Sci.* **2021**, *11*, 2029. https://doi.org/10.3390/app11052029

Academic Editor: Florent Ravelet

Received: 30 January2021
Accepted: 22 February 2021
Published: 25 February 2021

1. Introduction

The phenomenon of a structure impacting on the water generally implies large forces due to the massive fluid volume displaced in a small time. The duration of the slamming event is in the order of milliseconds. These loads might damage the structure or, because of their short duration, excite a dynamic response of the local structure of the hull and cause vibrations. Non-symmetric impacts might further introduce a tilting moment and influence the impact dynamics and structural stability.

The first analytical solution to solve the impact dynamics of rigid bodies entering the water was presented by Von Karman [1], who developed a formula capable of predicting the maximum force acting on a rigid body entering the water, to make a stress analysis on the members connecting the fuselage with the floats of a seaplane. Many analytical methods have been proposed to extend Wagner's method to different shapes (e.g., [2–7]) and most of them are very effective in predicting the water entry of simple-shaped structures impacting the surface with pure vertical velocity. Some of these solutions are even capable of accounting for oblique impacts (e.g., [8–12]). It is reported that particular conditions of entry velocity, deadrise angle, and tilt angle might lead the fluid to detach at the wedge apex, introducing difficulties in evaluating the pressure at the interface by analytical formulations. A typical water entry event that largely deviates from a symmetric case is ditching [13–15], where fluid detachment and cavitation arise if the advancing velocity is sufficiently large. Another area of interest for asymmetric impacts is the analysis of the response of planing hulls during the maneuvering operation since, depending on the conditions, restoring or capsizing moments can take place [16]. Xu [17] defined two types of asymmetric impact. Type A flow is the one when there is small asymmetry and the flow moves outward along the contour on both sides of the vertex. Type B flow occurs when there is large asymmetry and the flow detaches from the body contour at the vertex on one side. Chekin [18] concluded that there was only one unique combination of wedge angle and impact angle at which no separation of flow from the vertex would occur. For a given wedge shape and impact direction, any other impact angle would force separation. Defining U as the horizontal velocity and V as the vertical velocity, the advancing ratio U/V at which the flow separation appears is less for bodies of larger deadrise angles. For small asymmetry impacts, the cavity flow during the water-entry is limited to a very small region. Furthermore, the flow that separates from the apex quickly re-attaches to the wedge. A symmetric body impacting with horizontal velocity will produce a flow similar to asymmetric impact with only vertical velocity when a tilting motion is not permitted. Judge et al. [8] performed experiments on wedges where asymmetry and horizontal impact velocity were present and compared the results with an analytical solution, showing good agreement for low angles of asymmetry and small advancing ratios.

Xu et al. [17] observed that the pressure near the tip of asymmetric wedges might attain negative values with respect to the gauge pressure, and hence be lower than the ambient pressure. They also observed that the peak pressure increases with the horizontal velocity, and with it, an increase in the wave elevation on the upstream side and a decrease on the downstream one. However, the simulations are based on the assumption of attached flow. In this work, we also consider the eventual detachment at the wedge tip, similarly to that described in [19] but utilizing a different numerical scheme. Semenov [20] studied the effect of the horizontal component of the entry velocity for various wedge orientations. Results show that certain configurations of the impact might induce a negative pressure along the whole wetted length of the downstream side of the wedge, leading to flow separation at the wedge apex. Semenov and coauthors [21–23] derived an analytical solution for the asymmetric/oblique water entry of a wedge which does not assume flow separation from the wedge vertex. It was shown that at some wedge orientations, the total force acting on the wedge side with the larger deadrise angle may become zero, a condition corresponding to flow separation.

Experimental evidence of the onset of cavity formation at high asymmetry ratios has been provided by Shams et al. [24], who studied asymmetric water impacts experimentally through particle image velocimetry.

In this manuscript, we initially define the problem of asymmetric water impacts in Section 2 and we detail the numerical model utilized for the analysis in Section 3. Asymmetric water entry events are then divided into categories and studied separately in Sections 4–6, to finally draw some final remarks in Section 7.

2. Problem Statement: Water Entry of Asymmetric Wedges

In the most general case, wedges enter the water with combined vertical and horizontal velocity components, whose ratio defines the advancing ratio ϵ, and are not symmetric due to an initial tilt angle γ. As mentioned in the introduction, this kind of impact might introduce ventilation into the fluid flow [8] due to fluid detachment at the wedge tip. A sketch of the problem is shown in Figure 1: a wedge with nominal deadrise angle β enters the water at the velocity V_0 and is rotated by a tilt angle γ with respect to the water surface.

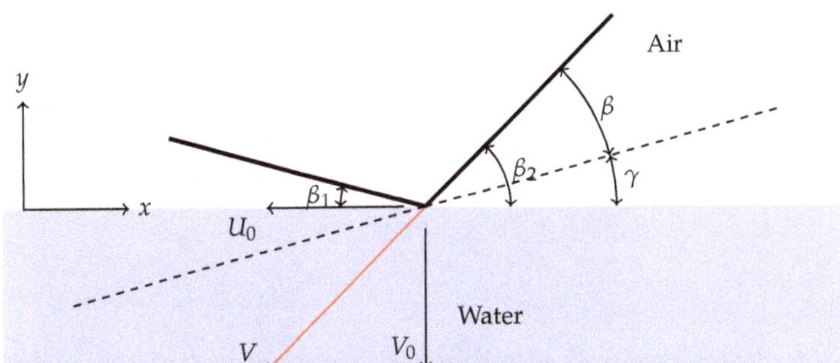

Figure 1. Sketch of the asymmetric impact of a wedge.

In the following, the analysis will always refer to wedges designed as symmetric structures with nominal deadrise angle β but rotated by a tilt angle γ, hence presenting different deadrise angles on the two sides ($\beta_1 = \beta - \gamma$ and $\beta_2 = \beta + \gamma$). We also define the advancing ratio ϵ as the ratio between the horizontal velocity U_0 and the vertical velocity V_0. Utilizing such definitions, we here classify the water entry event into three main categories:

- $\gamma = 0$ and $\epsilon \neq 0$—symmetric wedge with an horizontal velocity component;
- $\gamma \neq 0$ and $\epsilon = 0$ —asymmetric wedge with pure vertical velocity;
- ϵ, $\gamma \neq 0$—asymmetric wedge with an horizontal velocity component.

In the case of vertical impact, fluid detachment is prone to happen only if the condition $\beta + \gamma \geq \pi/2$ is met. In the presence of a horizontal velocity component, it is reported in the literature [20] that the flow separation from the wedge vertex may occur when $\pi - \gamma_{\text{inf}} < \beta_L$, being β_L is the leeward-side deadrise angle, and γ_{inf} the water-entry angle, as in these cases, the pressure near the tip might become negative [25]. Although this phenomenon is well known, it is usually neglected due to the difficulties in treating the fluid detachment. This work instead focuses on those cases where the fluid detaches at the wedge vertex.

3. Coupled FEM/SPH Numerical Model

In this study, we assume that cavitation at the wedge vertex does not occur and flow separation depends on the flow characteristics enabling ventilation. Gravity is always neglected, while the entry velocity and tilt motion either follow the actual impact dynamics or is set to a constant value, for in the latter case the solution is self-similar in time, hence the results might be normalized by the entry depth. In the case of full fluid–structure interaction simulations (hence modelling the whole impact dynamics) the solution is not self-similar in time not only because of the role of acceleration but also because the tilt motion modifies the deadrise angle. As a result, the jet of water piling up over the counter-clock side thickens, while in the opposite side, the water is pushed by the rotating wedge and the free surface profile near the wedge deforms significantly.

The 2D numerical model , presented in Figure 2, is based on the SPH solutor available within the commercial software LS-DYNA. In all the numerical solutions, the fluid is modelled by SPH particles with a constant diameter of 0.25 mm covering a region 0.8 m wide and 0.3 m deep. Non-reflecting boundaries define the bounding box. The fluid is modelled through an equation of state following the Gruneisen model and the fluid particles are solved utilizing the SPH-enhanced fluid formulation (`FORM = 16`). Both wedge's sides are 0.3 m long and are modelled as a rigid shell composed by 100 elements on each side, leading to an average of four SPH particles that make contact with a single element in the wet region. Such a strategy allows to smooth out the marked pressure fluctuations otherwise attained at the fluid–structure interface. Deadrise angle, initial rotation, and entry velocities are parametrically varied, while the wedge mass is 100 kg per unit width.

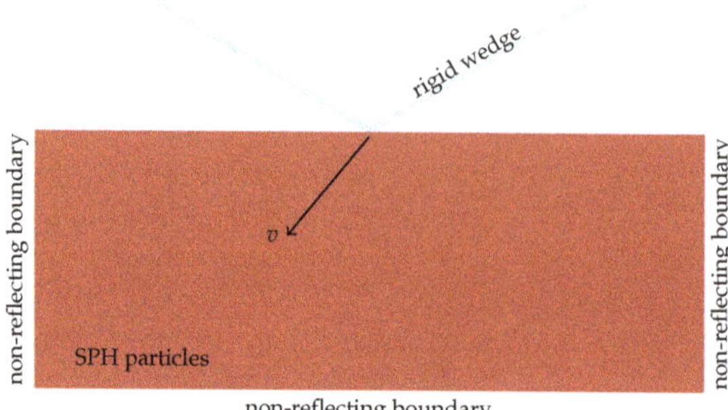

Figure 2. Two-dimensional numerical model with highlighted boundary and initial conditions.

The numerical results will be compared against established analytical solutions. It has already been shown that the numerical scheme utilized within this work is in line with the analytical predictions provided by Wagner's model for the case of pure vertical velocity [26,27] which, in turn, was shown to provide results in line with experimental results conducted on rigid wedges [28–31]. In Wagner's expanding plate model, which has been formulated for symmetric impacts only, the pressure distribution along the wedge edge is given by:

$$\frac{p}{\rho} = \dot{V}\sqrt{r^2 - x^2} + \frac{V\, r\, \dot{r}}{\sqrt{r^2(t) - x^2}} - \frac{1}{2}\frac{V^2\, x^2}{r^2 - x^2} \tag{1}$$

while the Modified Logvinovich Model (MLM), before flow separation, reads [32]

$$\frac{p}{\rho} = \frac{1}{2}V^2\left[\frac{\dot{r}}{V}\frac{r(t)}{\sqrt{r^2(t) - x^2}} - \cos^2\beta\frac{r^2(t)}{cr^2(t) - x^2} - \sin^2\beta\right] \tag{2}$$

where $0 < x < r$ defines the wet portion of the edge. Within this scheme, x is an abscissa oriented horizontally, i.e., following the direction of the free surface, and r is a function of the entry depth ξ and equals $\frac{\pi}{2}\frac{\xi}{\tan\beta}$. In the case of constant entry velocity, the horizontal projection of the wet length reads $r = \frac{\pi}{2}\frac{V\,t}{\tan\beta}$. MLM can be also utilized to study asymmetric impacts. However, a simplified method considering an equivalent deadrise angle equal to the average of the left- and the right-side deadrise angles is sufficiently accurate to predict the overall impact dynamics [32]. Within this numerical scheme, the distribution of the non-dimensional hydrodynamic pressure $\bar{p}$, which is given by $\frac{2p\tan\beta}{\rho V^2}$, over the horizontal abscissa x is given by

$$\bar{p}(x) = \frac{\pi}{\sqrt{1 - \left(\frac{x}{r}\right)^2}} - \tan\beta\frac{\left(\frac{x}{r}\right)^2}{1 - \left(\frac{x}{r}\right)^2} \tag{3}$$

The normalized pressure distribution for varying deadrise angle is shown in Figure 3.

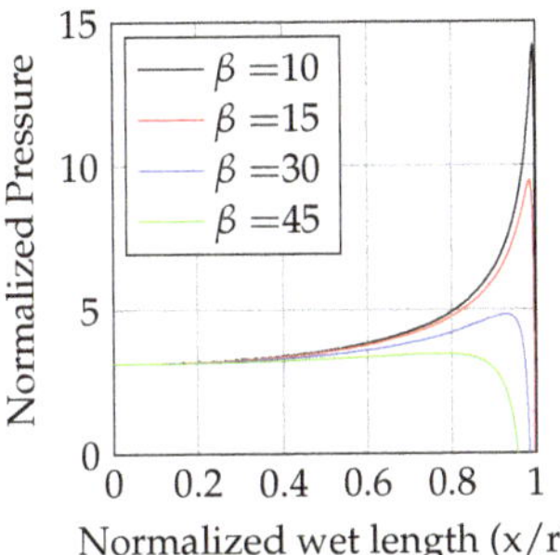

Figure 3. Evolution of the normalized pressure along the wetted edge for variable deadrise angles.

Following Wagner's solution, the vertical force (per unit width) of the wedge can then be estimated as follows:

$$F(t) = 2\int_0^1 \bar{p}dx\,\frac{\rho V^2}{2\,\tan\beta} r\cos\beta \quad \text{or} \quad F(t) = \bar{F}\,\frac{\pi\,\rho V^3\,t}{2\,\tan\beta}\frac{\cos\beta}{\tan\beta} \tag{4}$$

This shows that the ratio between the vertical force and the entry depth, which in the case of constant velocity equals Vt, is

$$\frac{F(t)}{V\,t} = \bar{F}\frac{\pi\rho V^2}{2\,\tan\beta}\frac{\cos\beta}{\tan\beta} \tag{5}$$

which is time-independent, being a function of the deadrise angle only. Here, vertical force, horizontal force, and tilting moment will be normalized as

$$\bar{F}_{x,y} = F_{x,y}\frac{2\tan^2\beta}{\cos(\beta)\pi\rho V_y^3 t} \quad \text{and} \quad \bar{M} = M\frac{2\tan^2\beta}{\cos(\beta)\pi\rho V_y^2 V_x t^2}. \tag{6}$$

4. Symmetric Wedges with Horizontal Velocity Component—$\gamma = 0$ and $\epsilon \neq 0$

In this section, we investigate the effect of a horizontal velocity component on the pressure distribution over a rigid wedge. At first, we concentrate on rigid wedges with constant velocity and null tilt motion, since such approximation leads to a self-similar solution and the impact time can be taken out of the analysis.

As a general result, the horizontal velocity component alters the fluid motion beneath the wedge up to the point a cavity is formed. Indeed, a threshold value of it exists for the onset of cavity formation. Figure 4 shows three images of the cavity predicted by the numerical solutions on a wedge with $\beta = 25°$ and varying advancing ratios ϵ. Notably, fluid detachment is attained as ϵ exceeds 2 and the void region increases with the advancing ratio.

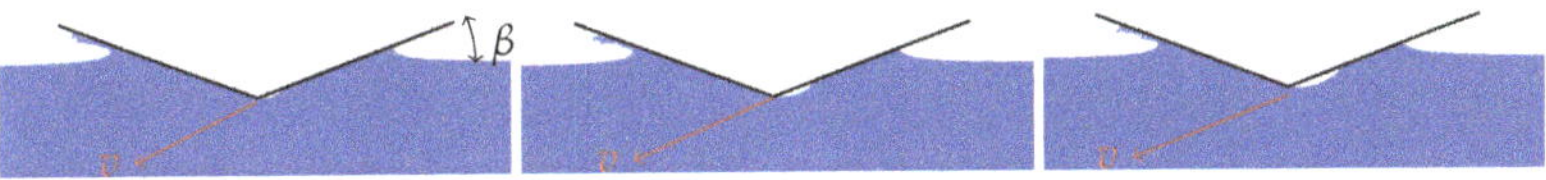

Figure 4. Fluid detachment at the vertex of a $\beta = 25°$ wedge for varying advancing ratio $\epsilon = \frac{6}{3}$, $\frac{7}{3}$and $\frac{8}{3}$ (left to right).

Indeed, pressure drops to zero in the dry region, the extension of which increases with the advancing ratio ϵ. Figure 5 details the predicted pressure distribution over a symmetric wedge with deadrise angle $\beta = 30°$ entering the water with varying advancing ratios ϵ. These examples report wedges running at a constant speed and the tilt motion is inhibited. Each graph reports the normalized pressure over the normalized wet portion of the wedge at various instants of the impact. Notably, the solution remains self-similar for all the advancing ratios considered, as the normalized pressure and void region are not influenced by the entry depth. Interestingly, the leeward wedge side remains completely dry as ϵ exceeds 2.

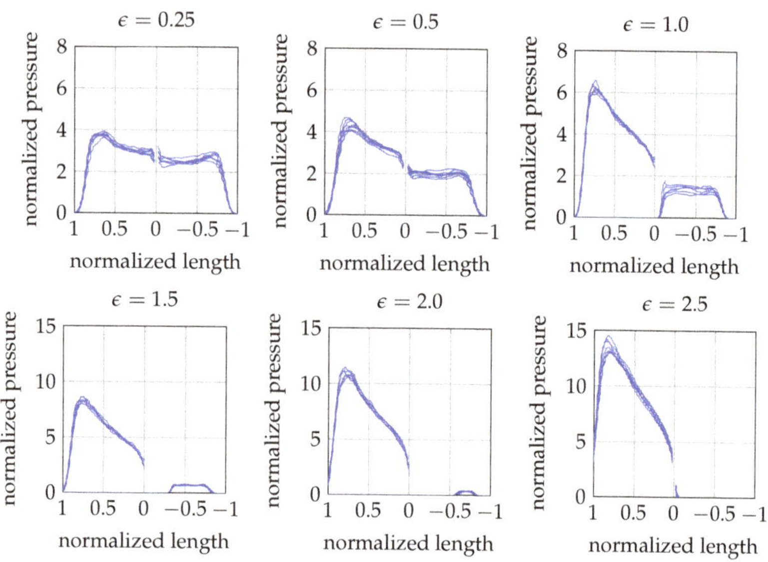

Figure 5. Normalized pressure versus wet portion of a wedge ($\beta = 30°$) at various instants of the impact for varying ϵ.

The self-similarity of the results allows for the effect of the advancing ratio on the pressure distribution along the body to be compared. Figure 6 reports results similar to the ones presented in Figure 5, but stacked on a single graph to ease their comparison. Results about $\beta = 20°$ and $\beta = 30°$ wedges with varying advancing ratios are presented. In the case of pure vertical impact, such deadrise angles experience a hydrodynamic load that is almost constant over the entire wet surface, as also predicted by the Wagner's solution reported in Figure 3. However, when a horizontal velocity component is introduced, the pressure on the right side (that is, the windward side) is found to increase almost linearly moving from the vertex to the free edge, while it remains constant on the leeward side, except for the region experiencing fluid detachment. The graph reports a single solution for each case due to the self-similarity of the results. In the case of $\beta = 30°$, fluid detachment initiates when ϵ becomes greater than 1 and the magnitude of the void region with respect to the overall wet length increases with the advancing ratio, to eventually remain fully dry when ϵ is greater than 2. Notably, for both the deadrise angles considered, the cavity incipient condition is met as soon as $\beta + \arctan\epsilon \approx 1.3$, while the cavity is completely formed as soon as $\beta + \arctan\epsilon \approx \pi/2$, a result which matches with the analytical predictions [25], validating the capability of the numerical model to correctly capture the cavity formation event. The pressure on the windward side is found to increase with the horizontal component while it decreases on the opposite side. At the higher advancing ratio considered in this study ($\epsilon = 2.5$), the left side never makes contact with the fluid. Figure 6 also highlights, with particular reference to the $\beta = 30$ case, how the pile-up on the windward side of the wedge increases with the advancing ratio, while it decreases on the leeward side. These results should be ascribed to the increasing horizontal component of the velocity.

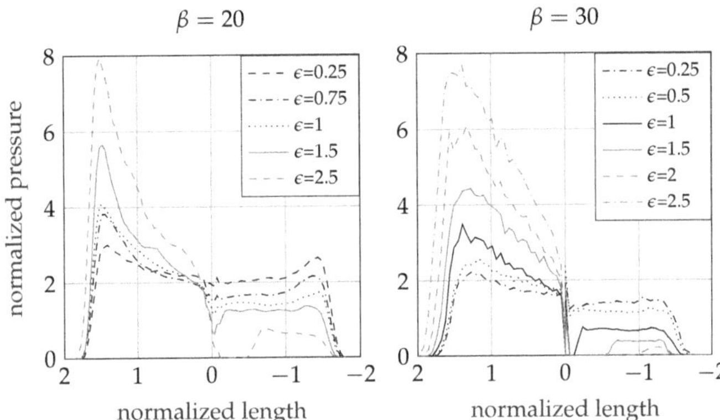

Figure 6. Normalized pressure versus normalized wetted surface on a symmetric wedge with $\beta = 30°$ (**left**) and $\beta = 20°$ (**right**), entering the water with different advancing ratios $\epsilon = V_x/V_y$.

Having imposed a constant entry velocity further allows normalizing the impact forces and the tilting moment, whose magnitude remain constant during the whole impact event. The normalized results for all cases considered are collected in Figure 7. The normalized horizontal force F_x is found to linearly increase with the advancing ratio, as the horizontal velocity component increases with it—while both the vertical force F_y and the momentum M_z remain constant (hence proportional to the vertical force only)—to eventually diverge as ϵ exceeds 2, which is the full cavity generation condition.

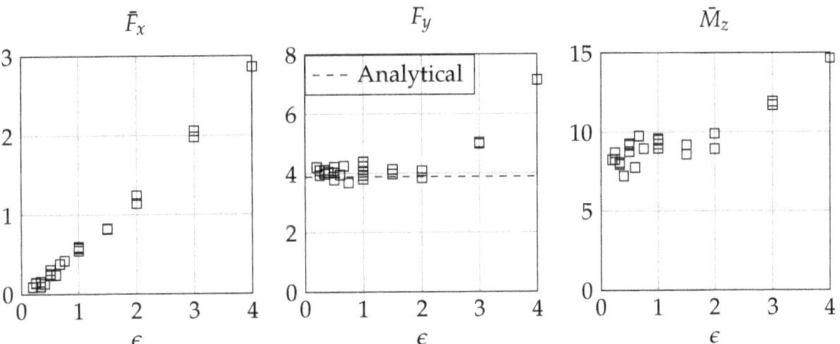

Figure 7. Normalized horizontal force, vertical force, and momentum versus advancing ratio ϵ.

5. Asymmetric Wedges with Pure Vertical Velocity—$\gamma \neq 0$ and $\epsilon = 0$

This section firstly proposes to adapt simple analytical formulations to investigate the impact dynamics of asymmetric wedges. Later, the hydrodynamic pressure evaluated numerically will be compared against Wagner's solution to assess its validity in asymmetric water impacts.

5.1. Impact Dynamics of Asymmetric Wedges

The impact dynamics of the asymmetric wedge impacting with pure vertical velocity will be here approximated through Wagner's formula, by adjusting the added mass term to account for the different deadrise angles, hence the wet length, of the two sides of the wedge. The mass of the expanding plate in Wagner's approach is here approximated by

$$m = \rho \frac{\pi}{2} \left[\frac{1}{2} \left(\frac{\xi}{\tan(\beta_0 + \gamma)} + \frac{\xi}{\tan(\beta_0 - \gamma)} \right) \right]^2 \tag{7}$$

And the impact dynamics is estimated utilizing Equation (4) by substituting the term $\tan(\beta)$ with

$$\tan(\beta) = 2\frac{\tan(\beta_0 + \gamma) \cdot \tan(\beta_0 - \gamma)}{\tan(\beta_0 + \gamma) + \tan(\beta_0 - \gamma)} \quad (8)$$

The comparison between analytical predictions and the numerical results shows a fairly good match during the initial stage of the water entry, to slightly diverge after the peak acceleration is reached. Such a difference should be ascribed to the tilt motion which modifies the deadrise angle over time, an effect that is not accounted for in the analytical model. As an example, Figures 8–11 show the comparison between numerical results and analytical predictions of the impact dynamics of a $\beta = 25°$ wedge varying tilt angle and entry velocity.

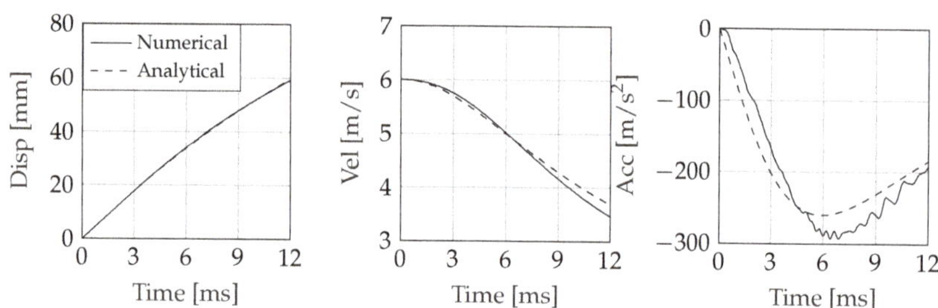

Figure 8. Impact dynamics of an asymmetric wedge entering the water at 6 m/s. Deadrise angle $\beta = 25°$, tilt angle 5°.

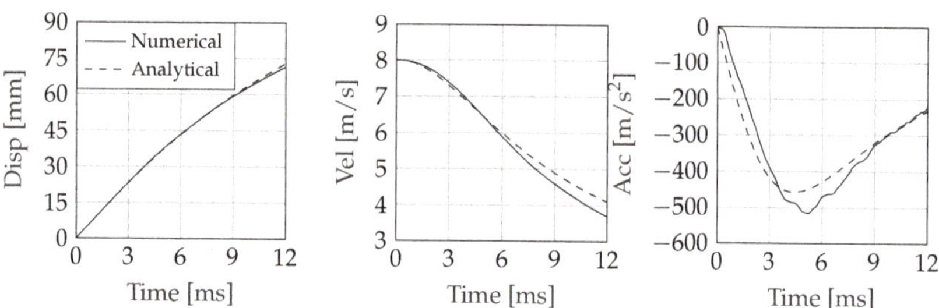

Figure 9. Impact dynamics of an asymmetric wedge entering the water at 8 m/s. Deadrise angle $\beta = 25°$, tilt angle 5°.

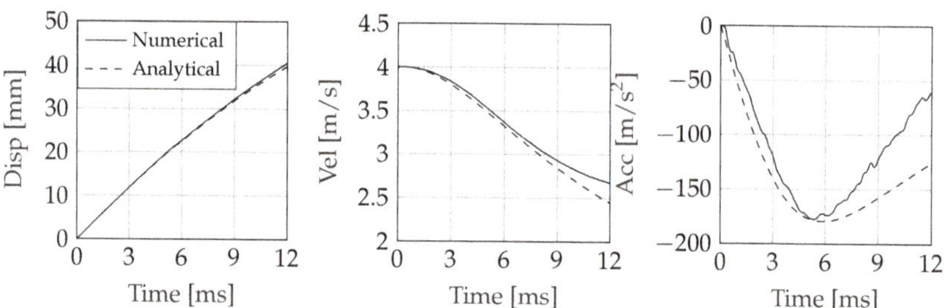

Figure 10. Impact dynamics of an asymmetric wedge entering the water at 4 m/s. Deadrise angle $\beta = 25°$, tilt angle 15°.

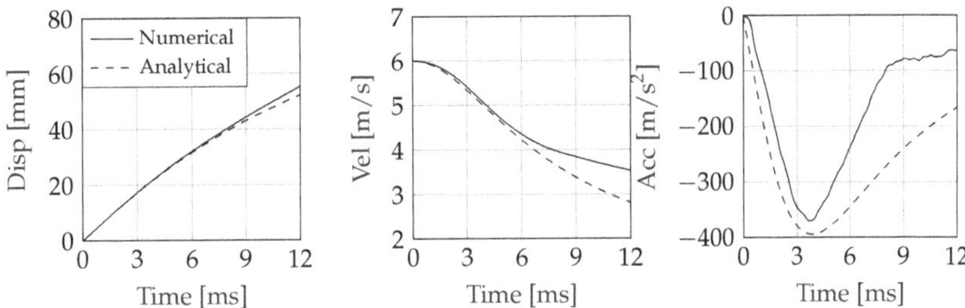

Figure 11. Impact dynamics of an asymmetric wedge entering the water at 6 m/s. Deadrise angle $\beta = 25°$, tilt angle 15°.

5.2. Hydrodynamic Pressure

It has been found that the impact dynamics during the initial stages of the water entry process can be effectively predicted by a simple modification of the expanding plate model. To evaluate the effect of the tilt angle on the pressure distribution, we now concentrate on the hydrodynamic pressure at the interface and compare the results against Wagner's predictions.

Figure 12 shows the effect of the tilt angle on the pressure distribution for the case of a $\beta = 30°$ wedge with constant entry velocity and null tilt motion. In such a condition, the solution is found to remain self-similar in time and the trend of the hydrodynamic pressure follows Wagner's prediction on each side of the wedge. The pileup coefficient [33] is not influenced by the initial tilt angle and it remains approximately constant to $\pi/2$, matching with the analytical value predicted for symmetric wedges.

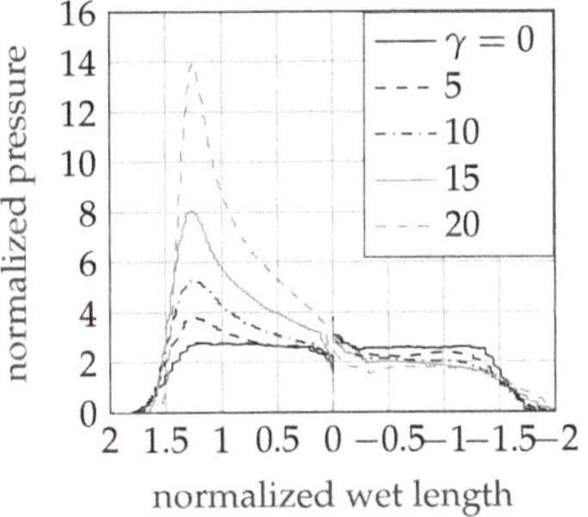

Figure 12. Effect of the initial tilt angle on the pressure distribution on a $\beta = 30°$ wedge entering the water with pure vertical velocity for varying initial tilt angle.

Indeed, if the velocity is not prescribed a priori, the simulation loses the self-similarity and the solution changes over time. Figure 13 reports, as an example, the time evolution of the pressure at the fluid–structure interface of a wedge with deadrise angle $\beta = 25°$ entering the water with an initial tilt angle of 5° and initial velocity of 6 m/s. The hydrodynamic pressure compares well during the initial stage of the impact but soon diverges in both sides, to eventually switch the side experiencing the higher pressure. As already mentioned before, such a difference should be ascribed to the tilt motion, which is not taken into account in the analytical model. Please note that the sudden pressure jump at the keel, the magnitude of which increases with ϵ, is a graphical artefact introduced by the normalization of the results.

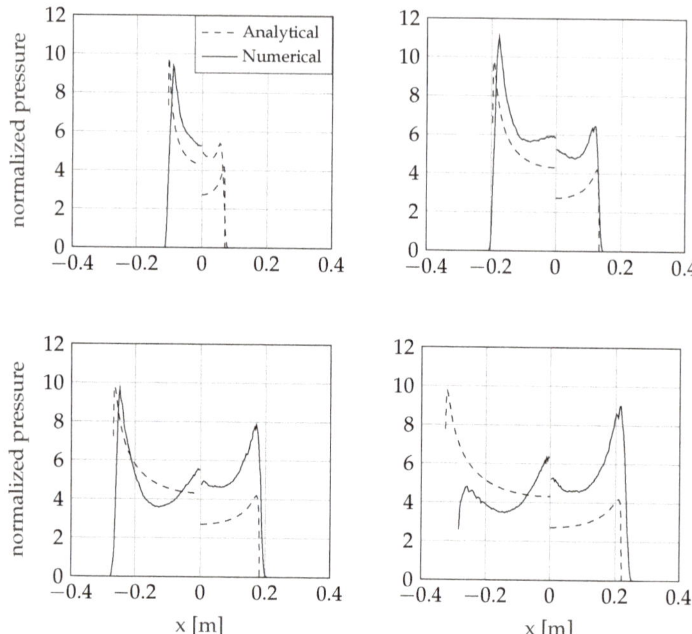

Figure 13. Asymmetric wedge entering the water at 6 m/s with pure vertical velocity. Deadrise angle $\beta = 25°$, tilt angle 5°. Pressure distribution at 4, 8, 12, and 16 ms from the impact.

The presented results show that, on one side, an eventual asymmetry during the water impact does not have a remarkable effect on the impact dynamics and on the maximum impact force, since the peak force is attained in the early stage of the impact, prior to a substantial tilt motion. On the other side, the tilt motion is found to largely alter the hydrodynamic pressure distribution at the larger entry depths.

The next section presents some numerical solutions for asymmetric wedges entering the water with combined vertical and horizontal velocity to investigate the combined influence of tilt angle and horizontal component.

6. Asymmetric Wedges with Horizontal Velocity Component—ϵ, $\gamma \neq 0$

In this section, asymmetric wedges with combined vertical and horizontal impact velocities are studied, which represents a combination of the impact conditions studied in Sections 4 and 5. The analysis considers only the cases where the advancing ratio is in the same direction of the tilt angle (left direction in Figure 1, and counter-clock wise rotation). Advancing ratio, deadrise angle, and tilt angle have been parametrically varied in the numerical analysis, but only some reference cases are reported here for brevity.

As an example, two cases representative of incipient cavity formation are presented: Figure 14 compares the numerical results and analytical predictions of the pressure over a 25° wedge with an initial tilt angle of 15° entering the water with vertical velocity $v_y = 4$ m/s and horizontal velocity $v_x = 2$ m/s. Figure 15 shows a wedge with a deadrise angle of 25° and tilt angle of 5° entering the water with vertical velocity $v_y = 4$ m/s and horizontal velocity $v_x = 4$ m/s (ϵ=1).

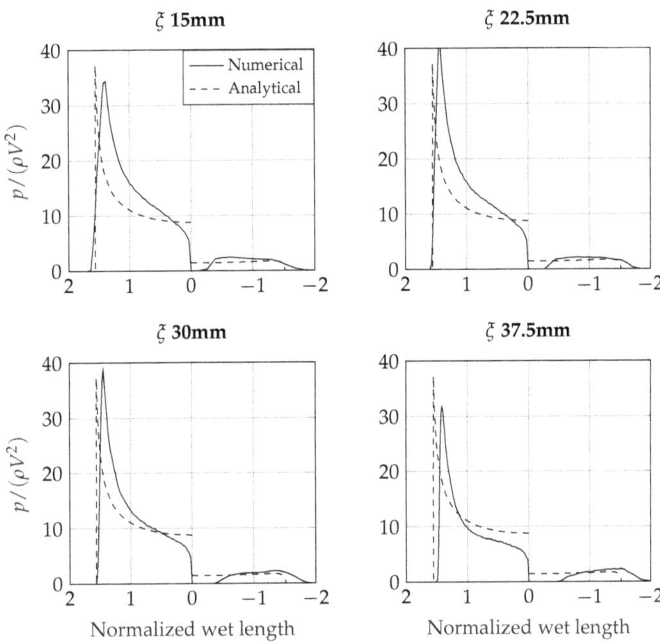

Figure 14. Asymmetric wedge entering the water at 4 m/s (y) + 2 m/s (x). Deadrise angle $\beta = 25^\circ$, tilt angle 15°.

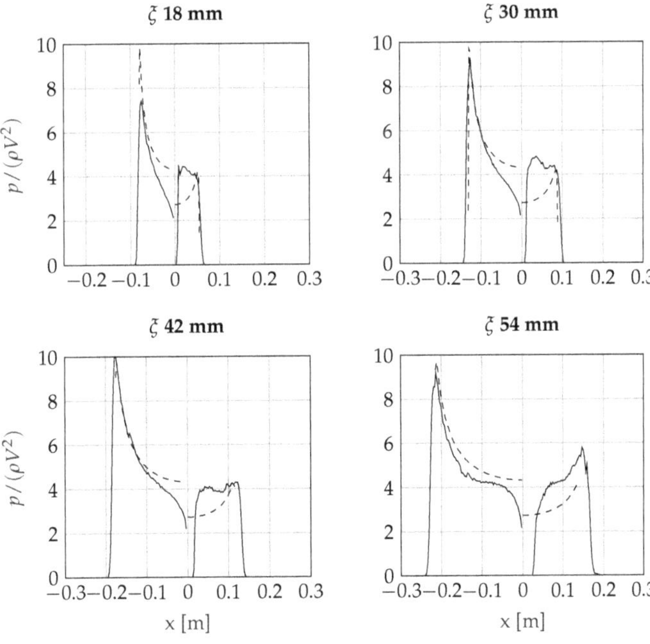

Figure 15. Asymmetric wedge entering the water at 4 m/s (y) + 4 m/s (x). Deadrise angle $\beta = 25^\circ$, tilt angle 5°.

In the case shown in Figure 15, the high horizontal velocity component induces the fluid to detach from the wedge tip, as visible in Figure 16, reporting the fluid velocity. Since air is not considered in the numerical model, the numerical solution might not represent the real behaviour of the fluid. Nevertheless, the results show that even if there is a region where the pressure drops to zero, the maximum pressure compared well with Wagner's model, even if the overall shape changed: the maximum pressure on the leeward face is now located where the fluid reattach to the wedge rather than in the fluid jet.

Figure 16. Velocity field in the fluid. Asymmetric wedge entering the water with combined horizontal (4 m/s) and vertical (4 m/s) velocity. Deadrise angle $\beta = 25°$, tilt angle 5°.

Overall, the results show that the combination of leeward deadrise angle and advancing ratio resulting in incipient cavity formation is $\beta + \arctan \epsilon > 1.2$, and full cavity formation is attained when $\beta + \arctan \epsilon > \pi/2$. These results are in line with the ones obtained for the symmetric wedges with horizontal velocity component presented in Section 4.

7. Conclusions

This work demonstrates that the proposed numerical solution scheme based on a coupled FEM–SPH approach allows for prediction of the hydrodynamic pressure during asymmetric water entry of rigid bodies and also in the case of cavity formation at the wedge vertex. The following three cases were considered: symmetric wedges entering the water with a velocity vector with components normal and tangential to the fluid surface, asymmetric wedges falling vertically, and, finally, the combination of asymmetry and a generic velocity vector. In the first scenario, the results show that the horizontal component of the velocity influences the fluid motion beneath the wedge and that a threshold value of the ratio between the horizontal and the vertical velocity components exists, leading to the onset of the cavity formation. This threshold depends on the deadrise angle. Cavity formation has an influence on the impact forces and on the tilting moment that increases significantly for advancing ratios above the threshold. On the other side, in pure vertical entry events, it is found that an asymmetry in terms of the initial tilt angle does not affect the maximum hydrodynamic load. This is very important because, as a consequence, the impact dynamics can be effectively predicted by simply introducing the tilt angle in the analytical formulations developed for symmetric impacts. However, the tilt motion alters the hydrodynamic pressure distribution at the larger entry depths. Finally, the combination of the two conditions shows that the maximum pressure is comparable to Wagner's solution, even if the pressure distribution is very different and presents zero pressure zones. As in the symmetric case, the cavity formation presents a threshold, which, in this case, is a function of the deadrise angle and the advancing ratio. As a further step, the proposed numerical scheme can be coupled with FEM to determine the effect of asymmetry on the hydro-elastic behaviour of flexible structures.

Author Contributions: R.P. contributed to carrying out the simulations and the data analysis. He is the main writer of the paper. G.M. contributed to the redaction of the manuscript and was responsible for the supervision. All authors have read and agreed to the published version of the manuscript.

Funding: This research received no external funding.

Institutional Review Board Statement: Not Applicable.

Informed Consent Statement: Not Applicable.

Data Availability Statement: The data presented in this study could be available on request from the corresponding author.

Conflicts of Interest: The authors declare no conflict of interest.

References

1. Von Karman, T. The Impact on Seaplane Floats, during Landing. NACA-TN-321, October 1929. Available online: https://ntrs.nasa.gov/citations/19930081174 (accessed on 30 January 2021).
2. Nikfarjam, M.; Koto, J.; Yaakob, O.B.; Seif, M.S.; Aref, A. Pressure Distribution at Water Entry of a Symmetrical Wedge. *J. Ocean Mech. Aerosp.* **2014**, *12*, 13–17.
3. Wu, G.X.; Sun, H.; He, Y.S. Numerical simulation and experimental study of water entry of a wedge in free fall motion. *J. Fluids Struct.* **2004**, *19*, 277–289. [CrossRef]
4. Dobrovol'skaya, Z.N. On some problems of similarity flow of fluid with a free surface. *J. Fluid Mech.* **2006**, *36*, 805. [CrossRef]
5. Korobkin, A.A. Second-order Wagner theory of wave impact. *J. Eng. Math.* **2006**, *58*, 121–139. [CrossRef]
6. Mei, X.; Liu, Y.; Yue, D.K.P. On the water impact of general two-dimensional sections. *Appl. Ocean. Res.* **1999**, *21*, 1–15. [CrossRef]
7. Battistin, D.; Iafrati, A. Hydrodynamic loads during water entry of two-dimensional and axisymmetric bodies. *J. Fluids Struct.* **2003**, *17*, 643–664. [CrossRef]
8. Judge, C.; Troesch, A.; Perlin, M. Initial water impact of a wedge at vertical and oblique angles. *J. Eng. Math.* **2004**, *48*, 279–303. [CrossRef]
9. Korobkin, A.A. Inclined entry of a blunt profile into an ideal fluid. *Fluid Dyn.* **1988**, *23*, 443–447. [CrossRef]
10. Seif, M.S.; Mousaviraad, S.M.; Sadathosseini, S.H. The effect of asymmetric water entry on the hydrodynamic impact. *Int. J. Eng. Trans. A Basics* **2004**, *17*, 205–212.
11. Algarín, R.; Tascón, O. Hydrodynamic Modeling of Planing Boats with Asymmetry and Steady Condition. In Proceedings of the 9th International Conference on High Performance Marine Vehicles (HIPER 11), Naples, Italy, 18–19 September 2011; Volume 2, pp. 1–9.
12. Truong, T.; Repalle, N.; Pistani, F.; Thiagarajan, K. An experimental study of slamming impact during forced water entry. In Proceedings of the 17th Australasian Fluid Mechanics Conference, Auckland, New Zealand, 5–9 December 2010.
13. Spinosa, E.; Iafrati, A. Experimental investigation of the fluid-structure interaction during the water impact of thin aluminium plates at high horizontal speed. *Int. J. Impact Eng.* **2021**, *147*, 103673. [CrossRef]
14. Iafrati, A.; Grizzi, S. Cavitation and ventilation modalities during ditching. *Phys. Fluids* **2019**, *31*, 052101. [CrossRef]
15. Iafrati, A.; Grizzi, S.; Olivieri, F. Experimental Investigation of Fluid–Structure Interaction Phenomena During Aircraft Ditching. *AIAA J.* **2020**, 1–14. [CrossRef]
16. Riccardi, G.; Iafrati, A. Water impact of an asymmetric floating wedge. *J. Eng. Math.* **2004**, *49*, 19–39. [CrossRef]
17. Xu, G.; Duan, W.; Wu, G. Numerical simulation of oblique water entry of an asymmetrical wedge. *Ocean. Eng.* **2008**, *35*, 1597–1603. [CrossRef]
18. Chekin, B.S. The entry of a wedge into an incompressible fluid. *J. Appl. Math. Mech.* **1989**, *53*, 300–307. [CrossRef]
19. Krastev, V.K.; Facci, A.L.; Ubertini, S. Asymmetric water impact of a two dimensional wedge: A systematic numerical study with transition to ventilating flow conditions. *Ocean Eng.* **2018**, *147*, 386–398. [CrossRef]
20. Semenov, Y.A.; Yoon, B.S. Onset of flow separation for the oblique water impact of a wedge. *Phys. Fluids* **2009**, *21*, 112103. [CrossRef]
21. Yu, B.; Semenov, Y.A.; Iafrati, A. On the nonlinear water entry problem of asymmetric wedges. *J. Fluid Mech.* **2006**, *547*, 231. [CrossRef]
22. Semenov, Y.A.; Wu, G.X. Asymmetric impact between liquid and solid wedges. *Proc. R. Soc. A Math. Phys. Eng. Sci.* **2013**, *469*, 20120203. [CrossRef]
23. Goman, O.G.; Semenov, Y.A. Oblique entry of a wedge into an ideal incompressible fluid. *Fluid Dyn.* **2007**, *42*, 581–590. [CrossRef]
24. Shams, A.; Jalalisendi, M.; Porfiri, M. Experiments on the water entry of asymmetric wedges using particle image velocimetry. *Phys. Fluids* **2015**, *27*, 027103. [CrossRef]
25. Xu, G.D.; Duan, W.Y.; Wu, G.X. Simulation of water entry of a wedge through free fall in three degrees of freedom. *Proc. R. Soc. A Math. Phys. Eng. Sci.* **2010**, *466*, 2219–2239. [CrossRef]
26. Panciroli, R.; Abrate, S.; Minak, G.; Zucchelli, A. Hydroelasticity in water-entry problems: Comparison between experimental and SPH results. *Compos. Struct.* **2012**, *94*, 532–539. [CrossRef]

27. Panciroli, R. Water entry of flexible wedges: Some issues on the FSI phenomena. *Appl. Ocean Res.* **2013**, *39*, 72–74. [CrossRef]
28. Panciroli, R.; Porfiri, M. Evaluation of the pressure field on a rigid body entering a quiescent fluid through particle image velocimetry. *Exp. Fluids* **2013**, *54*, 1630. [CrossRef]
29. Panciroli, R.; Abrate, S.; Minak, G. Dynamic response of flexible wedges entering the water. *Compos. Struct.* **2013**, *99*, 163–171. [CrossRef]
30. Panciroli, R.; Porfiri, M. Analysis of hydroelastic slamming through particle image velocimetry. *J. Sound Vib.* **2015**, *347*, 63–78. [CrossRef]
31. Jalalisendi, M.; Zhao, S.; Porfiri, M. Shallow water entry: Modeling and experiments. *J. Eng. Math.* **2016**. [CrossRef]
32. Qin, H.; Zhao, L.; Shen, J. A modified Logvinovich model for hydrodynamic loads on an asymmetric wedge entering water with a roll motion. *J. Mar. Sci. Appl.* **2011**, *10*, 184–189. [CrossRef]
33. Panciroli, R.; Pagliaroli, T.; Minak, G. On Air-Cavity Formation during Water Entry of Flexible Wedges. *J. Mar. Sci. Eng.* **2018**, *6*, 155. [CrossRef]

Article

Analysis of Bulb Turbine Hydrofoil Cavitation

Andrej Podnar, Marko Hočevar, Lovrenc Novak and Matevž Dular *

Faculty of Mechanical Engineering, University of Ljubljana, Aškerčeva 6, 1000 Ljubljana, Slovenia; andrej.podnar@gmail.com (A.P.); marko.hocevar@fs.uni-lj.si (M.H.); lovrenc.novak@fs.uni-lj.si (L.N.)
* Correspondence: matevz.dular@fs.uni-lj.si; Tel.: +386-1-477-1453

Featured Application: Bulb water turbine blades with improved cavitation properties.

Abstract: The influence of a bulb runner blade hydrofoil shape on flow characteristics around the blade was studied. Experimental work was performed on a bulb turbine measuring station and a single hydrofoil in a cavitating tunnel. In the cavitation tunnel, flow visualization was performed on the hydrofoil's suction side. Cavitation structures were observed for several cavitation numbers. Cavitation was less intense on the modified hydrofoil than on the original hydrofoil, delaying the cavitation onset by several tenths in cavitation number. The results of the visualization in the cavitation tunnel show that modifying the existing hydrofoil design parameters played a key role in reducing the cavitation inception and development, as well as the size of the cavitation structures. A regression model was produced for cavitation cloud length. The results of the regression model show that cavitation length is dependent on Reynolds's number and the cavitation number. The coefficients of determination for both the existing and modified hydrofoils were reasonably high, with R^2 values above 0.95. The results of the cavitation length regression model also confirm that the modified hydrofoil exhibits improved the cavitation properties.

Keywords: hydrofoil; bulb turbine; bulb turbine runner; cavitation; flow visualization; cavitation tunnel; regression model

Citation: Podnar, A.; Hočevar, M.; Novak, L.; Dular, M. Analysis of Bulb Turbine Hydrofoil Cavitation. *Appl. Sci.* **2021**, *11*, 2639. https://doi.org/10.3390/app11062639

Academic Editor: Florent Ravelet

Received: 12 February 2021
Accepted: 10 March 2021
Published: 16 March 2021

1. Introduction

Historically, hydropower was used as a source of baseload generation for electric power, but now it is increasingly used for grid balancing. Among dispatchable renewable energy sources, hydropower plants are the most often used [1], achieving up to 15 s response times [2] for 90% of the total power response. The future potential of hydropower is shifting towards grid regulation tasks. Including nondispatchable renewable energy sources, e.g., inter alia wind and photovoltaics, and market deregulation, creates the requirement for rapid power generation fluctuations and power increase in the afternoon [3–5]. Any new hydraulic turbine design should therefore enable this type of operation and reach high efficiency and stable operation outside its optimal operating interval. Besides stable continuous turbine operation away from the best efficiency point, new designs will also need to function sufficiently in transient operation. Bulb turbines face challenges for such operation; designers will have to overcome the detrimental effects that flow fluctuations and cavitation have on turbine operation, while still ensuring adequate performance over a broad interval of operating points and in transient operation [3].

At the best efficiency point, an optimal combination of the runner frequency of the rotation, head, and discharge is available. At such an operating point, flow separation is usually low, and no cavitation is observed. For operation away from the optimum efficiency point, bulb turbines, as compared e.g., to Francis turbines [4], feature double regulation and may operate in a wide interval of operating conditions with reasonably high efficiency. Although all hydraulic turbine types are constantly improved, the gains in efficiency are in the per mille range and the cavitation in hydraulic turbines remains an issue. The cavitation in the hydraulic turbines [4,6,7] decreases efficiency and increases vibrations and erosion, thereby reducing the service life.

The procedures optimizing the performance of turbine hydraulics are most often based on the use of so-called inverse design methods [8] and computational fluid dynamics (CFD) [9].

Inverse design methods have been in use for decades, with the first 3D methods emerging around 30 years ago. These methods are based on the a priori specification of the pressure or blade loading distribution. Such distribution along the runner blade in the radial and meridional directions facilitates the calculation of the runner blade's angle. This reduces the computational CFD work required when insufficient information on the runner geometry is known in advance. Zangeneh [10] proposed a three-dimensional design method for radial and mixed turbines for the compressible case. Today his method is still often used for designing radial turbine machinery. Zhu et al. performed inverse design optimization of the reversible pump-turbine [11], while also considering cavitation characteristics. Leguizamón and Avellan [12] very recently provided an open-source implementation of an inverse Francis turbine design, together with a validation study confirming the approach's appropriateness. Cavitation is not often studied using inverse design methods; among the available studies, Daneshkah and Zangeneh [13] linked the blade loading and blade stacking to the efficiency and cavitation characteristics of a Francis runner, providing a valuable physical understanding and useful design guidelines.

The 3D design of the runners and the guide apparatus may be determined by performing consecutive CFD calculations, while constantly varying 3D geometry and design parameters [9,14]. Unfortunately, CFD alone is unable to establish an optimized 3D design. A 3D design study using CFD was performed by Xue et al. [15], considering the efficiency, stability of design, and cavitation of a pump turbine during pumping mode. A comprehensive hydraulic and parametric analysis of a Francis turbine runner was performed by Ma et al. [16]. They [16] performed a multipoint and multiobjective optimization design procedure, achieving efficiency improvements up to 0.91%. The optimization procedure included the 3D inverse design method, CFD analysis, and multipoint and multiobjective optimization technology.

Runner blades are central turbine elements, where energy is transferred from the fluid to the generator shaft. Blade sizes and angles, including hydrofoil selection, contribute to the blade design. The blade design is often a trade-off between efficiency, cavitation development, and cavitation erosion. A pressure difference between the suction and pressure sides is limited by cavitation, while at the same time the high pressure difference enables energy transfer from water to the generator. If the turbine is at full load and the water level is low in the reservoir, the pressure on the suction side may decrease and locally drop below the vapor pressure. Cavitation on the runner blades may appear in several locations. These are the turbine blade leading edge, the tip, the pivot, the gap between the blade tip and the discharge ring, the blade suction side close to the trailing edge, and the cavitation near the draft tube cone [7,17]. For the abovementioned cavitation types, the root causes are suboptimal runner blade design and hydrofoil selection for the selected operating conditions. Cavitation is extensively studied in hydraulic turbine machinery; among these, few studies have employed the visualization method [6,7].

On bulb turbines' blades, the initial cavitation most often appears at the location with the highest local speed i.e., at the blade tip [18]. We show an experimental analysis of a modification to the hydrofoil design of bulb turbine blades so as to resist cavitation during continuous operating conditions at high flows. The paper shows the cavitation characteristics of two bulb turbine hydrofoils. The improvement from the existing to modified hydrofoil is discussed based on visualization analysis in the cavitation channel. Furthermore, we show a comparison of the modified runner blades with the unmodified blades in terms of the turbine output. The paper shows marked improvement in cavitation occurrence between existing and modified hydrofoils, and a confirmed regression model of the void fraction.

2. Hydrofoil

Bulb turbine (Figure 1) blade hydrofoils are optimized for a mix of parameters, among which are maximum hydraulic efficiency, high output power, minimum turbine vibrations, high strength, reduced fatigue, and reduced cavitation phenomena. Reduced cavitation enables operators to reduce the service intervals required for cavitation damage repair and welding costs, as well as loss of money for nonoperation.

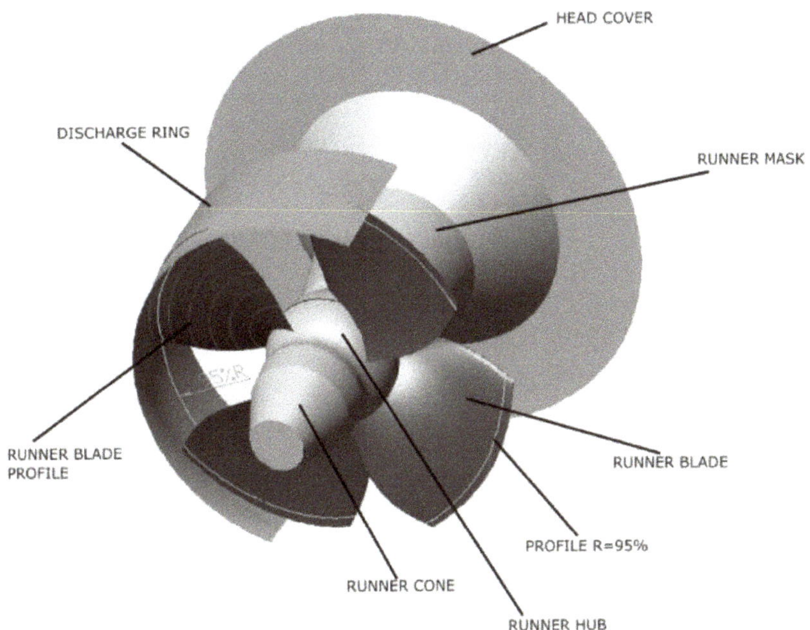

Figure 1. Bulb turbine runner.

In this study, we focus on a range of volumetric flows, which appear on a bulb turbine at a 25° blade angle (Figure 2). Here cavitation frequently occurs at approximately 15% higher volumetric flows than at the best efficiency point [6]. The 25° blade angle was chosen because it lies in the middle of the angle interval usually agreed between the producers and customers of bulb turbines. The blade angle 25° is often reinforced during performance or acceptance tests [18], because it corresponds to both common operating conditions and features increased probability for cavitation.

The bulb runner blade hydrofoil's shape determines the flow conditions in the runner. Well-shaped 3D blade hydrofoils reduce the inception and the growth of cavitation on the bulb turbine runner hydrofoils. Turbine hydrofoil profiles were used in designing the bulb runner blade's 3D shape, spanning from the runner hub toward the blade tip (Figure 2). For reliable 3D characterization of the turbine runner blade, and thus the entire bulb runner, several hydrofoils are usually selected. Later, a smooth surface is drawn over all hydrofoils to form the entire 3D geometry of the runner blade, further enabling CFD or experimental analysis.

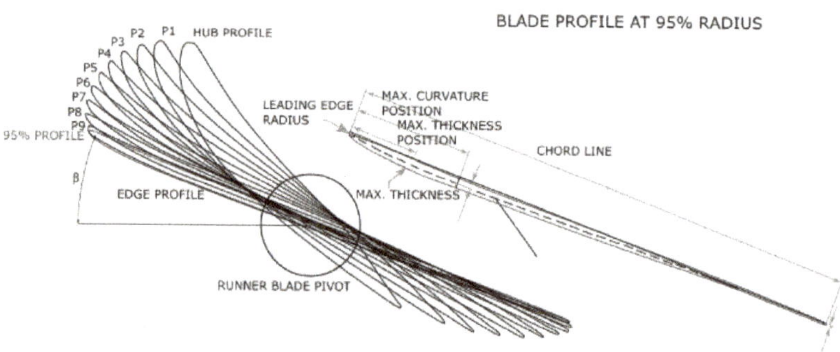

Figure 2. Selection of a 25° angle on a modified bulb turbine hydrofoil.

Although the bulb turbine blade can consist of many hydrofoils as explained above, in this paper we focus on a single hydrofoil. The most potential for energy conversion in bulb turbines is available near the tip, where tangential blade velocity and flow velocities are high. Due to high velocities, large pressure gradients, and the vicinity of the tip gap flow, this region is highly susceptible to cavitation occurrence and surface erosion. Here, several cavitation structures are usually found during acceptance tests. To focus on the cavitation properties of the hydrofoils, we have selected for further analysis the hydrofoil at 95% bulb turbine diameter D, measured from the runner rotation axis to the blade tip. The selected hydrofoil is thus located near the tip as shown in Figures 2 and 3. We made this selection on the basis of the energy transfer. Most of the energy is transferred where the turbine blade has a large diameter and the blade has a high surface area, and where the blade and flow velocities are high.

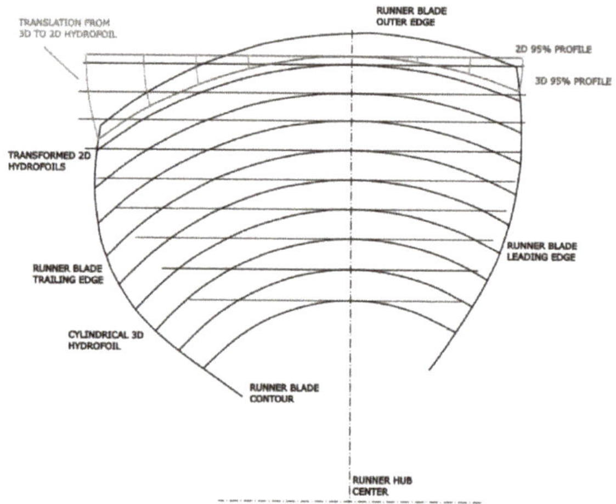

Figure 3. Bulb turbine blade design from individual 2D hydrofoils.

Hydrofoils have been studied extensively, including profile families such as NACA (among them 4–8 series), Göttingen, Munk, NPL, and others [19–21]. The available families of hydrofoils facilitate the turbine engineer's selection of the blade's hydrofoil to suit the required operating intervals and conditions, inter alia, for operation at high volumetric flows and runner blade angles. To enable operation in specific operating conditions, including cavitation properties, the turbine designer selects hydrofoil properties like leading edge radius, camber, location of maximum thickness, maximum curvature, and trailing edge design.

The locations of cavitation occurrence on bulb turbine blades are well documented in [22]. In the hill diagram, we will focus on the high discharge region. Here, the cavitation is found in four regions [22]: the blade suction side, blade pressure side, fillet, and tip gap. The flow properties in and near the gap are dominated by turbulence and may appear in the entire hill diagram. The hydrofoil properties that most affect the cavitation on the blade suction and pressure side are the leading edge radius, the location of maximum thickness, and the curvature; we may also assume that the blade angle in bulb turbines is always optimal due to the turbine's dual regulation. We want to design a leading edge such that the flow around it is smooth so as to prevent any flow separation. As discussed above, we will focus on the hydrofoil at 95% of the blade radius, where we assume that in addition to the mentioned leading edge, maximum thickness and curvature influence, the cavitation properties are also influenced also by the presence of the gap flow. At 95% of the blade radius, the flow properties (velocity and pressure) are more pronounced because of higher blade velocity than near the hub. The sharp leading edges may introduce flow separation [23] and deteriorate the flow properties further downstream from the leading edge. Downstream from the leading edge the flow is determined by the camber shape, the position of maximum curvature, maximum thickness, etc. These parameters, among others, determine the velocity distribution in the blade channel from the pressure to the suction side. Velocity is related to absolute pressure; an increase in flow velocity decreases the local pressure [7], thus, the velocity distribution affects the cavitation properties. The pressure distribution may cause local cavitation inception, and both can influence boundary layer separation.

The hydrofoils in the bulb or any other water turbines are the base elements of the turbine runner blades and are thus rotating. The bulb turbine features pressure distribution with the pressure normally decreasing from the leading edge towards the trailing edge. According to the bulb turbine's triangles of velocity, downstream from the hydrofoil the relative velocity increases and reaches its maximum near the trailing edge [7], and the situation is remarkably different in comparison with the single hydrofoil in the water tunnel. Furthermore, the pressure distribution around the hydrofoil in the cavitation tunnel cannot fully represent the pressure properties around the bulb turbine blade. The situation in the bulb turbines is therefore very complex and the flow properties around individual hydrofoils cannot be directly correlated around the bulb turbine runner blades. Nevertheless, we will assume [6] that bulb runner cavitation can be related to the cavitation analysis of a single hydrofoil in the water tunnel.

Hydrofoil Modifications

The turbine blade hydrofoils are the primary component of the turbine blades. To improve performance, the turbine hydrofoil profiles should be smoothly and consistently remodelled in a spatial 3D redesign of the entire bulb turbine runner blade. The new blade differs from the old in the following parameters: the leading-edge radius has increased, the maximum thickness location has moved forward, and the location of the maximum hydrofoil curvature has moved forward. The characteristics of both turbine hydrofoils are shown in Table 1.

Table 1. Characteristics of hydrofoils as used on bulb turbine.

Parameter	Modified	Unmodified
Runner diameter D_0	ø 350 mm	ø 350 mm
Number of runner blades Z	4	4
Location of reference hydrofoil R_h	95% D_0	95% D_0
Blade angle β_0	25°	25°
Blade chord line length at 95% R	180.7 mm	180.7 mm
Location of maximum thickness l_1	27.11 mm	72.28 mm
Ratio l_1/l	15%	40%
Location of maximum curvature l_2	54.2 mm	81.3 mm
Ratio l_2/l	30%	45%
Leading edge radius r_0	1.63 mm	1.27 mm

The modified hydrofoil as compared to the unmodified one features reduced absolute velocity in a section from the location of the maximum hydrofoil thickness and curvature to the trailing edge. The cavitation properties of the modified hydrofoil as compared to the unmodified one are in this section favorable. Because the points of maximum thickness and curvature for the modified hydrofoil profile were moved forward towards the leading edge, this section is relatively long.

The unmodified hydrofoil in a region from the leading edge to the location of the maximum thickness is very sensitive to the optimum angle of attack due to its sharp leading edge. Away from the optimum attack angle, the unmodified hydrofoil has disadvantageous cavitation characteristics in comparison with the modified one [6]. Contemporary water turbines often operate away from the best efficiency points due to the introduction of renewable energy sources and environmental changes. This trend will most probably increase in the future.

Two dimensional hydrofoils (Figure 4) were created from the 3D spatial bulb turbine blade hydrofoils (Figures 2 and 3). The 3D turbine blade hydrofoil was transformed from the cylindrical coordinate system of a turbine blade to a planar Cartesian x–y coordinate system as shown in Figure 3.

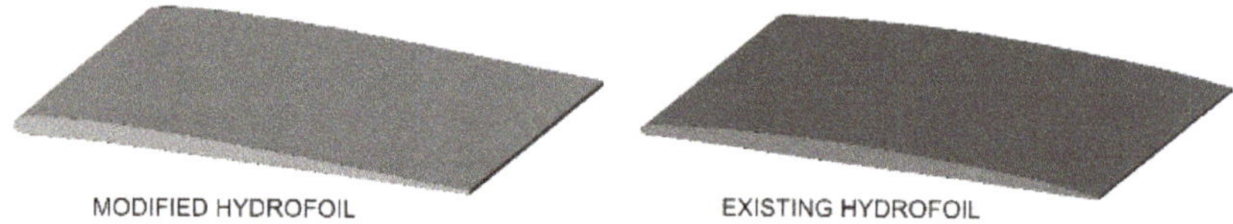

Figure 4. Hydrofoils used in the cavitation tunnel analysis.

As shown previously, both hydrofoils correspond to the runner blades' profiles at 95% R (Figure 3). The properties of both bulb runners are given in [4]. The turbine runners and both 2D hydrofoils were manufactured from brass using the same manufacturing techniques and both were given the same surface processing. The hydrofoils were later subjected to flow and image analysis in a cavitation tunnel and regression modelling.

The hydrofoil parameters for the experiment in the cavitation tunnel are shown in Table 2. The same manufacturing method as for the hydrofoils on the bulb turbine was used for the manufacture of the hydrofoils used in the cavitation tunnel.

Table 2. Hydrofoil parameters.

Parameter	Modified Hydrofoil	Existing Hydrofoil
Hydrofoil chord length l	60 mm	60 mm
Maximum hydrofoil thickness d_{max}	2.55 mm	2.55 mm
Maximum thickness position l_1	9 mm	24 mm
Maximum thickness position ratio l_1/l	15%	40%
Maximum hydrofoil curvature s_{max}	0.87 mm	0.87 mm
Maximum curvature position l_2	18 mm	27 mm
Maximum curvature position ratio l_2/l	30%	45%
Radius at the Leading edge r_0	0.54 mm	0.42 mm

3. Experiment

Experiments were performed on the turbine measuring station for both bulb turbine runners and in the cavitation tunnel for both hydrofoils. A turbine measuring station was used to determine efficiency for a single runner blade angle. In the cavitation tunnel, cavitation properties were measured for analogous operating points.

3.1. Turbine Measurements

Measurements of efficiency were performed on a low-head'axial and bulb turbines measuring station for low-head bulb turbines (Figure 5) in Kolektor Turboinštitut (Ljubljana, Slovenia). The measuring station was built according to standard [18] and was used for the model, witness, and acceptance tests. In agreement with standard [18], all measurements and data acquisitions were automatic. The nominal runner diameter of the test rig is 350 mm, and the recirculating flow is generated using two radial pumps with frequency regulation. The flow measurement was performed by using an electromagnetic flowmeter with a diameter of DN 400 and an accuracy of ±0.15%. The static pressure was measured using a differential pressure transducer (accuracy of ±0.025%), with the measuring taps located on the inlet and the outlet. Absolute suction pressure was measured using an absolute pressure transducer with an accuracy of ±0.025% to assure that the measuring station operated well above the incipient cavitation number. An electromagnetic incremental sensor was used to measure the rotational speed; the output was processed by a counter-timer module with an accuracy of ±0.01%. The shaft torque was measured on a horizontal shaft using a rotating torque transducer with an accuracy of ±0.01%. The friction torque was estimated by operation with air.

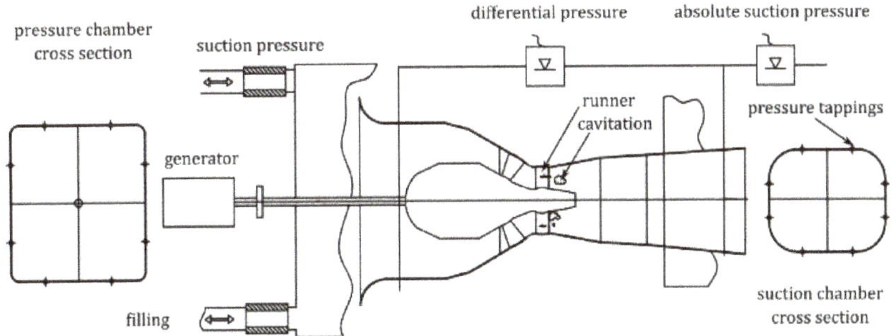

Figure 5. Turbine measuring station.

Both bulb turbine runners were tested under similar conditions at six operating points. The blade angle was cases set to 25° in all cases (Figure 6). The blade angle was set using a control template. For comparison, both runners operated at the same specific energy E (J/kg) and the corresponding nondimensional energy numbers ψ were recorded. E was determined from the total specific energy at the turbine inlet and the outlet cross-sections [7]. Energy numbers ψ were determined using the equation:

$$\psi = \frac{2 \cdot E}{\pi^2 \cdot N^2 \cdot D^2} \quad (1)$$

Here N (min^{-1}) is the rotational frequency, while D (m) is the discharge ring diameter of the bulb turbine. The turbine rotational frequency N (1000 min^{-1}) was the same for all measuring points. Both runners had the same nominal diameter D (m). The guide vane opening coefficient A_0 was the same for each pair of measuring points used in the comparison:

$$A_0 = \frac{a_v \cdot z}{D_v} \quad (2)$$

a_v (mm) is the guide vane's opening and z (/) is the number of guide vanes. D_v (mm) is the guide vane pivot diameter. Important manufacturing methods and characteristics were kept the same for reliable comparison between both runners. Control of runner tip gaps to the discharge ring and clearances of guide vanes to the guide vane pressure rings are very important for reliably estimating model turbine runners. Both runners were manufactured from the same material and using the same production methods and procedures. As such, for both runners, the tip gaps, guide vane clearances, and surface finishes were the same.

The measuring station was of a closed type. To maintain the ψ fixed, the main water circulation pump's frequency was controlled. The volumetric flow Q (m^3/s) represented in its nondimensional form as flow number φ [18] is:

$$\varphi = \frac{Q}{\pi^2/4 \cdot N \cdot D^3} \quad (3)$$

The results from a previous study show that the increase in efficiency amounts to around 1% for the same runner geometry. Most of the increase according to the hill diagram is due to the small shift of the operating points towards low flow numbers and higher efficiency [6].

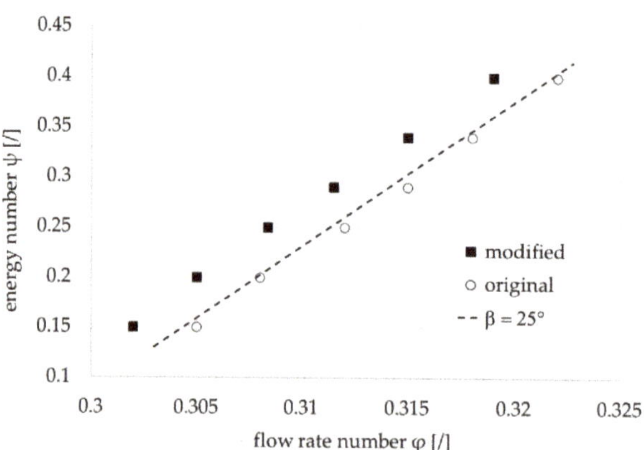

Figure 6. Turbine measuring points.

3.2. *Cavitation Tunnel Measurements*

The flow visualization measurements were taken on the cavitation measuring station. The approach to investigating the turbine blade flow properties in a separate tunnel experiment was similar to that used in [24]. A closed-type cavitation measuring station was used as shown in Figure 7. The flow was provided by a centrifugal pump driven by a frequency variable drive. The volumetric flow measurement was performed using an electromagnetic transducer ABB COPA-XL. Absolute pressure p_{abs} was measured on the suction side using an ABB 2600T 264NS pressure transducer. The temperature was measured by a Pt-100 4 wire temperature transducer, connected to the Agilent 34970A data acquisition unit. Valves in front of and behind the test section were fully open during experiments and they were only used during installation. Suction pressure was set using a vacuum pump, located on the top of a tank with a free water surface.

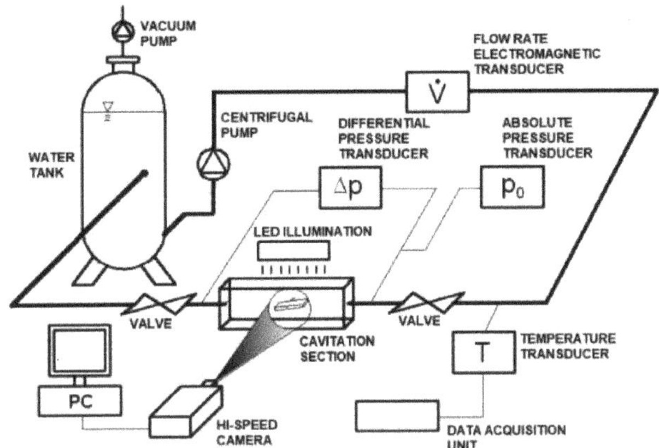

Figure 7. Cavitation measuring station with measuring and visualization equipment.

The cavitation tunnel test section cross-section was 50 cm × 100 cm. Hydrofoil width was 50 mm and chord length was l = 60 mm. The hydrofoil was installed at the angle of attack 8°.

A high-speed Photron SA-Z camera was used together with a Nikkor 105 mm 1:2.8 lens to visualize the cavitation in the cavitation section. The camera was mounted from the side. The camera was operated in a free-running mode. Images were recorded using PFV software in 8-bit BMP black and white mode at a resolution of 640 × 280. The framerate used was 25,000 frames/s. Shutter speed was equal to 10 μs. For each operating point, 2000 frames were recorded. For all experiments, all other camera settings were the same. A high intensity continuous LED illumination was provided by many CREE XLamp XM-L LED modules. LED modules were powered by an Instek DC power supply. Illumination was provided from the top of the test section above the hydrofoil on its suction side and arranged such that as few reflections as possible were present in the recorded sequences. Illumination intensity was not the same for all experiments; with the modified hydrofoil it was necessary to increase illumination intensity to visualize weaker cavitation structures. Keeping the illumination constant was impossible due to the low 8-bit depth of recorded images.

We have selected two dimensionless numbers to describe the cavitation in the cavitation tunnel, namely the Reynolds number Re and cavitation number σ. In Reynolds number Re

$$\mathrm{Re} = \frac{L \cdot v}{\nu} \tag{4}$$

L is hydrofoil length, v (m/s) is flow velocity in front of the hydrofoil and ν is the kinematic viscosity of water. To obtain the hydrofoil cavitation characteristics, the cavitation numbers σ were evaluated. Cavitation number σ was calculated according

$$\sigma = \frac{p_{abs} - p_v}{\rho \cdot v^2} \tag{5}$$

where p_v (N/m^2) is the water evaporation pressure and p_{abs} (N/m^2) is absolute pressure. Water evaporation pressure was estimate using the Equation

$$p_v = 10^{2.7862+0.0312 \cdot T_w - 0.000104 \cdot T_w^2} \tag{6}$$

where T_w (°C) is water temperature. Equation (6) provides for a maximum difference of 0.2%, with a negligible influence on the measurements.

The cavitation σ-η curve is used to estimate the cavitation properties of bulb turbines. The cavitation point for the highest cavitation numbers is measured at ambient pressure, while for subsequent cavitation points suction pressure is decreased. For the tunnel measurements the cavitation number σ was used to observe the cavitation's development on both hydrofoils. For each experiment, the absolute pressure p_{abs} was set as constant at the inlet of the cavitation tunnel. To decrease the cavitation number σ we increased the flow Q (m^3/s) and flow velocity. Several such operating points were measured for increasingly lower absolute pressure p_{abs} in the test section. This enabled comparison between both hydrofoils and among different absolute pressures.

The measuring points for both hydrofoils were selected according to cavitation structures sizes, shapes, and intensities. In general, cavitation intensity increased as cavitation number σ decreased. Instantaneous cavitation numbers σ were selected manually; we wanted to capture significant visible structural changes in cavitation. Such a procedure captured all important structures, starting with incipient cavitation on the leading edge, then sheet cavitation, cloud cavitation, and supercavitation across the entire hydrofoil [25,26]. Cavitation structure formation on both hydrofoils emerged at different cavitation numbers σ and, thus, the selection of cavitation number σ and corresponding measuring points for both hydrofoils was not the same.

3.3. Cavitation Image Analysis

Illuminated cavitation structures on the hydrofoils were recorded in a set of black and white images using the experimental setup from Section 3.2. The cavitation structures in the image have elevated grey level intensities, with vapor phase forming time and 2D space-dependent structures [22,27]. With computer-aided visualization of the process, the simultaneous time series of images were recorded. An assumption was made that the intensity of the illumination is proportional to the amount and intensity of the void fraction, being further proportional to the intensity of the cavitation structures in the observed regions [23]. This assumption is credible for cavitation intensity, low void fractions, and corresponding ratios of the intensity of the void fraction and grey level.

The time series of images from high-speed visualization in the cavitation tunnel were analyzed such that individual traveling bubbles were first filtered out. All bright objects in the individual image were deleted whenever they were very small. All other objects, like the attached cavitation or cavitation clouds, were retained. From these filtered images, the average grey level E_g (i,j,t) and standard deviation were calculated as explained below.

The grey level variable E_g (i,j,t) was acquired by processing intervals of 256 grey levels (8-bit camera image depth) from black (intensity = 0) to white (intensity = 255). The averaged grey level series were calculated in observation region. The standard deviation was estimated using the equation below:

$$S(i,j) = \sqrt{\frac{1}{T}\sum_{t=1}^{T}\left[E(i,j,t) - \langle E_g(i,j)\rangle\right]^2} \tag{7}$$

Here $\langle E_g (i,j)\rangle$ is time and spatially averaged grey level, which is proportional to the average void fraction of the cavitation structure's.

$$\langle E_g(i,j)\rangle = \frac{1}{T}\sum_{t=1}^{T} E(i,j,t) \tag{8}$$

The time interval t in Equations (7) and (8) is set from the duration and frequency of acquisition.

A sample of image analysis is shown in Figure 8 for the existing hydrofoil (p = 0.15 bar; Q = 61.7 m^3/h; σ = 2.27) and in Figure 9 for the modified hydrofoil (p = 0.15 bar; Q = 61.7 m^3/h; σ = 2.27). Cavitation length for the selected operating points was estimated to L = 22.74 mm for existing hydrofoil and L = 11.27 mm for modified hydrofoil. Such a method of estimating cavitation length was used for all operating points for both hydrofoils and later for regression cavitation length modelling.

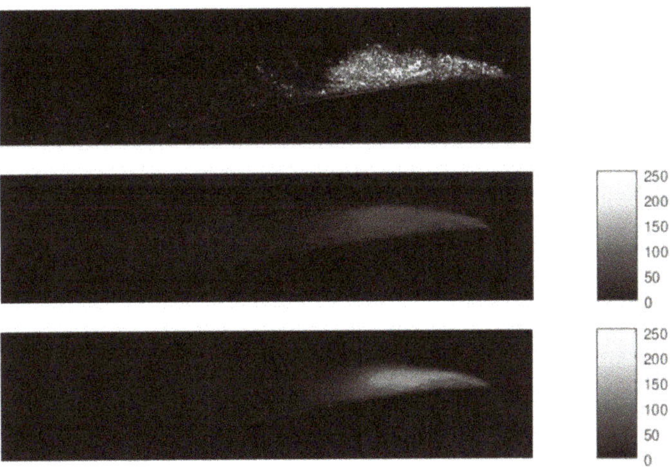

Figure 8. Original hydrofoil, a sample image of cavitation (**top**), spatially averaged grey level (**middle**), and standard deviation of grey level (**bottom**), operating point p = 0.15 bar, Q = 61.7 m^3/h, and σ = 2.27.

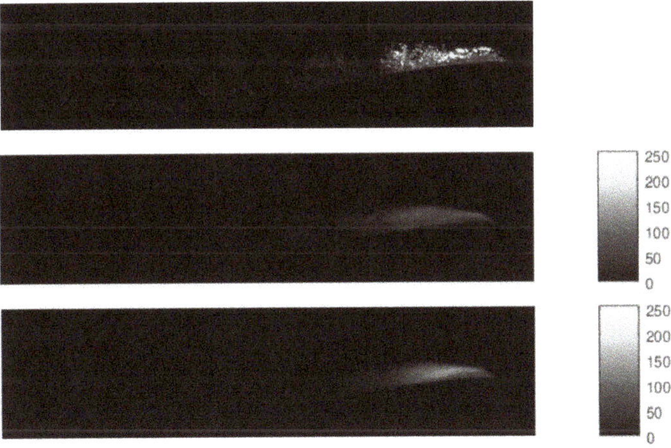

Figure 9. Modified hydrofoil, a sample image of cavitation (**top**), spatially averaged grey level (**middle**), and standard deviation of grey level (**bottom**), operating point p = 0.15 bar, Q = 61.7 m^3/h, and σ = 2.27.

3.4. Regression Cavitation Length Model

Cavitation properties are a consequence of flow parameters, which are best described using Reynolds number Re and cavitation number σ. We will show that cavitation length on a hydrofoil in a cavitation tunnel may be well predicted based on these two dimensionless numbers. A good correlation is expected between the measured and calculated cavitation lengths. For this, we will use the following power-law regression model for cavitation length L

$$L = \beta_0 \cdot Re^{\beta_1} \cdot s^{\beta_2} \quad . \tag{9}$$

Model parameters β_0, β_1, and β_2 were calculated using the minimum mean square error method.

4. Results and Discussion

We have shown in Figure 6, that two bulb turbine runners with differently shaped blade hydrofoils (Figure 4) show very similar pressure characteristics. Because both runners are different, their hill diagrams and characteristics, in general, cannot be compared. The analysis of the cavitation properties of both hydrofoils will be shown in the following section. As an attempt to provide the background for the bulb runner evaluation with a focus on cavitation, two differently shaped hydrofoils were tested in the cavitation channel. We observed cavitation phenomena as a result of the hydrofoil shape design. The main tool for analyzing the cavitation structures appearing on the hydrofoil surface was computer-aided visualization. From these cavitation images, the cavitation length was measured and used to make the regression model.

4.1. Cavitation Dependence on the Cavitation Number σ

The visual comparison and analysis of the two hydrofoils as a direct result of decreasing cavitation number σ is shown in Figures 10–13. Cavitation numbers from high to low were selected such that a marked difference in cavitation structures occurrence was observed. The selected images represent to the greatest extent possible the most typical conditions for selected cavitation number σ. The analysis was limited to the hydrofoil suction side. A decrease in the cavitation number increases the intensity and amount of the cavitation structures. We will demonstrate that shape modification is essential for cavitation to occur on hydrofoils. A modified hydrofoil will demonstrate a decrease in cavitation phenomena intensity in comparison to an existing hydrofoil. The main area of visualization is over the hydrofoil's leading edge and the entire suction side of the hydrofoil.

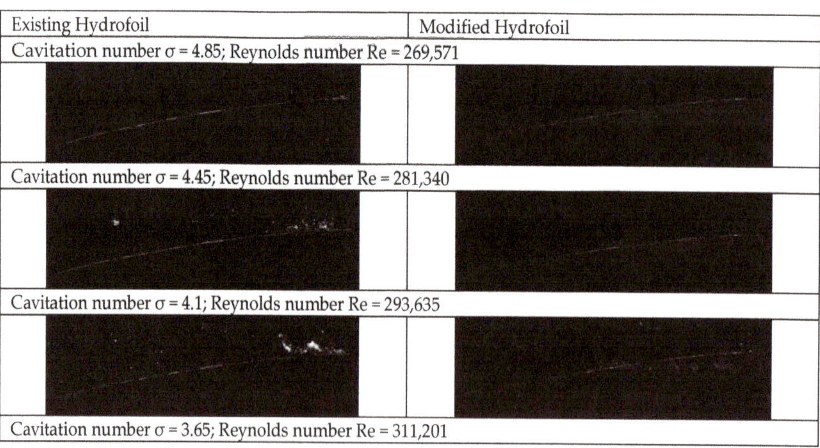

Figure 10. *Cont.*

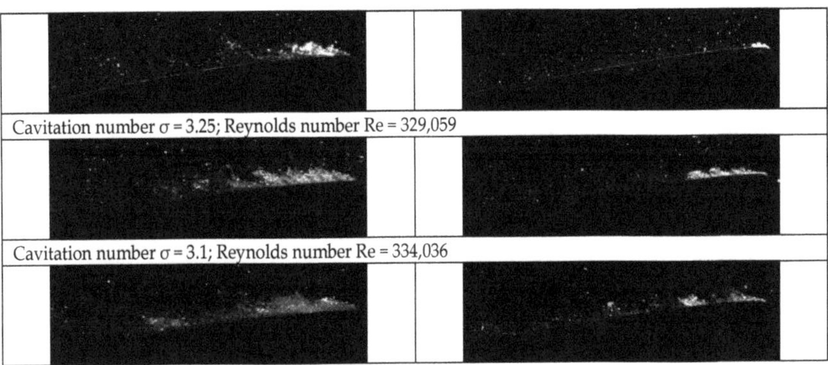

Figure 10. Visualization results of cavitation structures at $p_{abs} = 0.65$ bar.

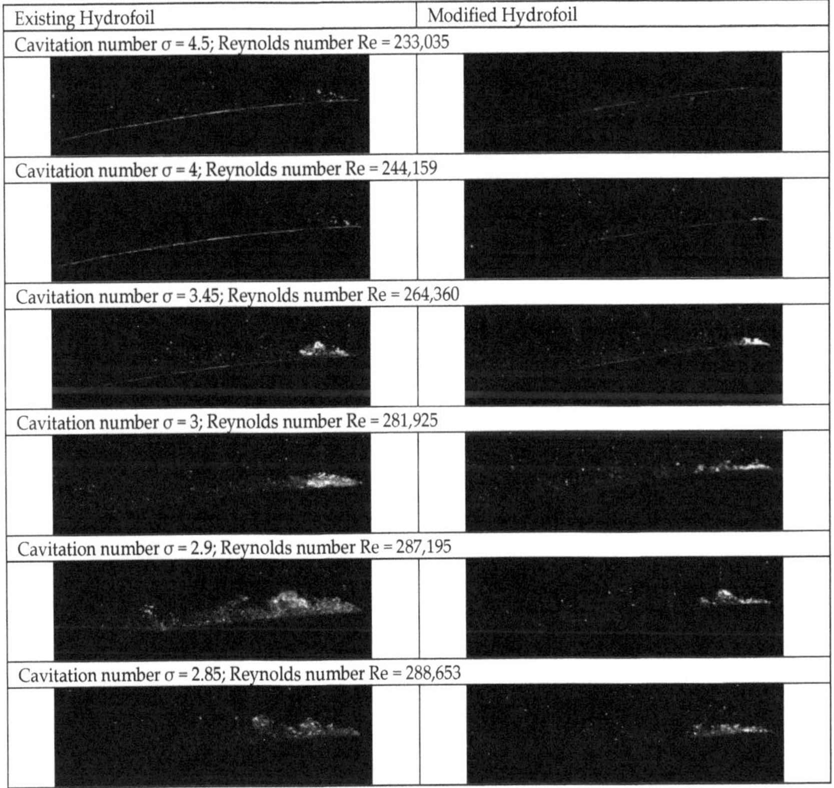

Figure 11. Visualization results of cavitation structures at $p_{abs} = 0.45$ bar.

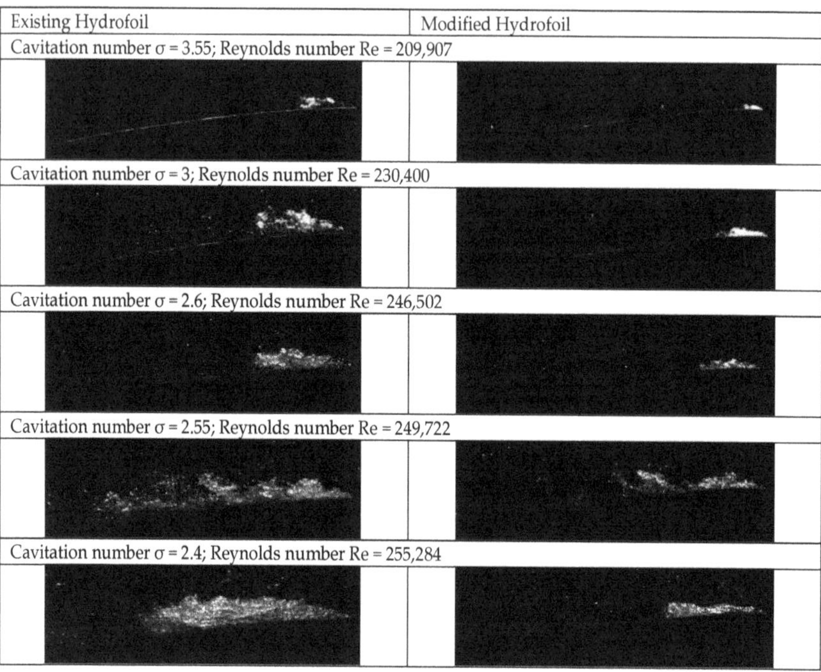

Figure 12. Visualization results of cavitation structures at p_{abs} = 0.30 bar.

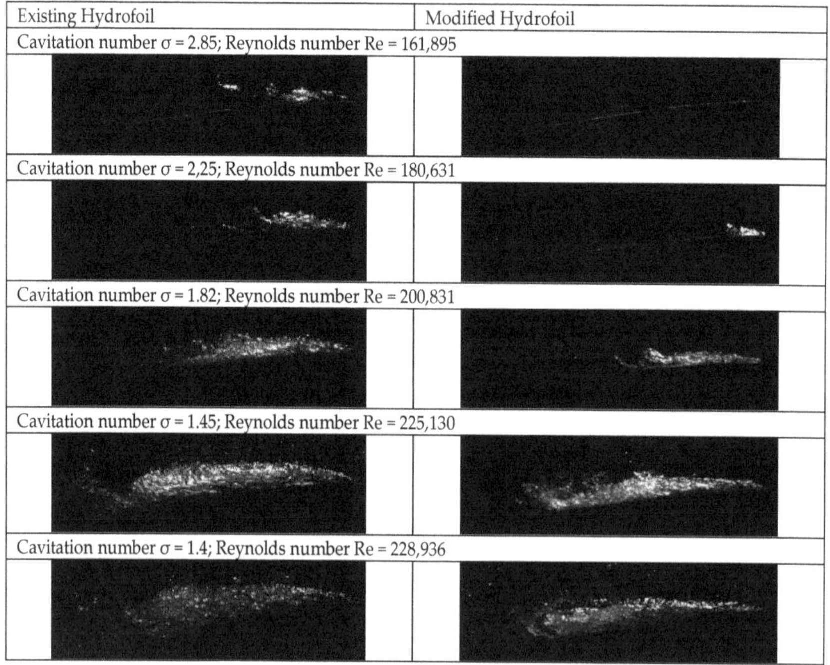

Figure 13. Visualization results of cavitation structures at p_{abs} = 0.15 bar.

Figures 10–13 are limited to the intervals of the cavitation number that we were able to achieve in the cavitation tunnel with relevant cavitation intensities.

The measurements at absolute pressure p_{abs} = 0.65 bar are shown in Figure 10. Cavitation structures (void fraction) start to appear on the modified hydrofoil at cavitation number around σ = 3.65. The cavitation starts behind the leading edge as an attached cavitation. With the further decrease of the cavitation number to around σ = 3.1 we observed an increase in the cavitation intensity and the quasiperiodic shedding of cavitation clouds. On the other hand, cavitation was observed on the original hydrofoil at cavitation numbers around σ = 4.45. At cavitation number σ = 3.65 we observed some cavitation cloud shedding, and at cavitation σ = 3.1 more than two-thirds of the hydrofoil was cavitating. At cavitation σ = 3.1, with the modified hydrofoil, the cavitation length was only 1/3 of the hydrofoil length. Since the hydrofoils in both cases were attached at the same angle of attack, we assume that the difference in cavitation occurrence is due to the shape and curvature of the leading edge. A sharp leading edge cuts through the flow such that it cannot follow the hydrofoil shape immediately downstream from the leading edge. The resulting flow separation induces cavitation inception due to the low absolute pressure and high velocity.

The measurements at absolute pressure p_{abs} = 0.45 bar are shown in Figure 11. The cavitation on the modified hydrofoil profile starts behind the leading edge as an attached cavitation around σ = 3.45, and p_{abs} = 0.65 bar. At cavitation σ = 3 we again observed cavitation cloud shedding as in Figure 10. The cavitation is much more intense for the case of the existing hydrofoil. We observed cavitation inception already at around σ = 4.5, while almost the entire hydrofoil was cavitating around σ = 3.

The measurements at the absolute pressure p_{abs}= 0.30 bar are shown in Figure 12. Measurements commenced around σ = 3.55; at this operating point the incipient cavitation is already present. Incipient cavitation here is present as an attached cavitation. For the modified hydrofoil, the attached cavitation formed only at around σ = 3; it covered up to around half of the hydrofoil length. With the original hydrofoil, the cavitation inception began around the same cavitation number (σ = 3.55) as for the modified hydrofoil, although with slightly higher intensity. A remarkable difference in the intensity of cavitation was observed at cavitation number σ = 3.0. Here, the cavitation cloud is much larger for the existing hydrofoil. Below the cavitation number σ = 2.6 most of the hydrofoil is covered by a thick cavitation cloud in the case of the existing hydrofoil.

The measurements at absolute pressure p_{abs} = 0.15 bar are shown in Figure 13. With modified hydrofoil, no cavitation was observed around cavitation number σ = 2.85, while with the modified hydrofoil the cavitation cloud at around σ = 2.25 covers up to one-third of the hydrofoil length. At around σ = 1.82 the cavitation cloud is again more intense than at cavitation number σ = 2.25. Below around cavitation number σ = 1.45 both hydrofoils are fully covered with supercavitation.

At cavitation numbers below around σ = 1.82 the large cavitation structures reduce the water passage area and increase the flow velocity in the cavitation tunnel at the selected flow. The results show favorable cavitation characteristics for the modified hydrofoil in comparison with the unmodified hydrofoil. We may also assume that the runner blade designed from such hydrofoils can perform well in the bulb turbine when operating under cavitation conditions.

4.2. Cavitation Length Model

Among several possible parameters of the cavitation flow properties, we have selected the cavitation cloud length for regression modelling. The results of the cavitation length model are shown in Table 3. Cavitation length dependence on the parameter β_0 is modest at around 6% in favor of existing hydrofoil. The modified hydrofoil's dependence on the parameter β_1 shows that the cavitation length of the modified hydrofoil is lower than that of the existing one. The most important improvement of the modified hydrofoil over the existing one is hidden in the dependence on cavitation number σ (achieving a value at parameter β_2 of −3.4, improving over the original −2.93). The results of the cavitation length model confirm the modified hydrofoil's superior cavitation characteristics over the existing one.

Table 3. Cavitation length model parameters for modelled cavitation length L.

	Parameter		
	β_0	β_1	β_2
Existing hydrofoil	2.81×10^{-8}	1.88	−2.93
Modified hydrofoil	2.99×10^{-8}	1.86	−3.40

The regression diagrams of the measured and modelled cavitation lengths are shown in Figures 14 and 15. The coefficients of determination for both the existing and modified hydrofoil profiles are reasonably good, with values above $R^2 > 0.95$. The high values of both coefficients of determination confirm the validity of the cavitation length model.

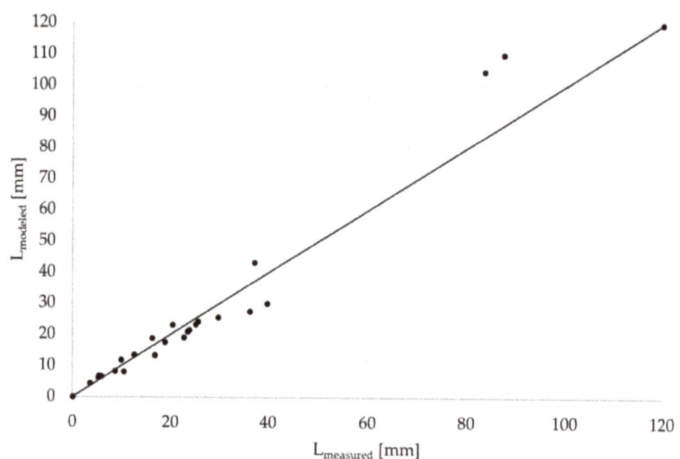

Figure 14. Cavitation model for existing hydrofoil, coefficient of determination $R^2 = 0.9586$.

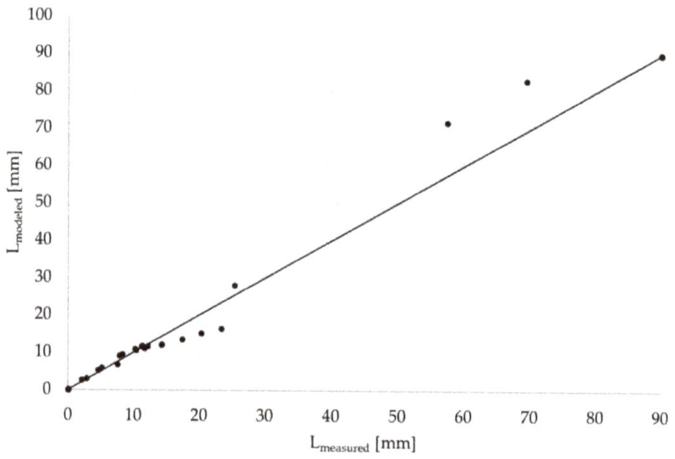

Figure 15. Cavitation model for the modified hydrofoil, coefficient of determination $R^2 = 0.9858$.

5. Conclusions

In future power grids, waterpower flexibility will be an important parameter for ensuring network stability. Water turbines will need to operate at wide flow intervals and in transient regimes, and also allow for fast response times. For this reason, among others, further research on cavitation is required.

The goal of this research was to demonstrate the relation between the hydrofoil shape and cavitation phenomena. Existing and modified hydrofoils were extracted from runner blades on the 95% runner blade radius and modified. The modifications were made on the hydrofoil's leading edge, as well as at the points of maximum thickness and curvature. The leading-edge radius was enlarged to avoid flow separation. Maximum thickness and curvature were moved in the direction of the leading edge to stabilize the flow around the hydrofoil towards the trailing edge. Both hydrofoils were implemented in a bulb turbine and measured under equal energy numbers. Measurements of both turbine runners were performed on the measuring station according to the standard for acceptance tests of water turbines.

The results of this study show improved cavitation characteristics. However, we show here only a single modification of the hydrofoil and the results only demonstrate the efficiency of this modification. Of particular benefit for future design would be a comprehensive study, providing for a hydrofoil and runner shape optimization algorithm, based on selection and later minimization of the objective functions, including the manufacture and measurements of a selection of hydrofoils and runners.

To evaluate the cavitation properties, the flow was visualized in the cavitation tunnel on the suction side. The modified hydrofoil exhibited improved cavitation properties and we chose to evaluate the intensity of the cavitation using the cavitation cloud length. A regression model prediction of the cavitation length shows that modified hydrofoil features improved cavitation characteristics in comparison with the existing hydrofoil.

The primary goal of this study was to provide for improved cavitation characteristics on the modified hydrofoil. Using the modified hydrofoil invariably changes runner characteristics and this resulted in operating points shifting to the left in the hill diagram. This leftward shift resulted in increased efficiency as shown in [6].

We believe that future requirements for water turbines will emphasize flexibility. For this reason, the future research on cavitation in water turbines will focus on the minimization of cavitation occurrence and erosion, while at the same time providing adequate energy output and efficiency.

Author Contributions: Conceptualization, A.P., M.H. and M.D.; methodology, A.P., M.H., L.N. and M.D.; software, A.P. and L.N.; validation, A.P., M.H., L.N. and M.D.; formal analysis, A.P., M.H. and L.N.; investigation, A.P., M.H., L.N. and M.D.; resources, A.P., M.H. and M.D.; data curation, A.P., M.H. and L.N.; writing—original draft preparation, A.P. and M.H.; writing—review and editing, A.P., M.H., L.N. and M.D.; visualization, A.P. and M.H.; supervision, M.H. and M.D.; project administration, A.P. and M.H.; funding acquisition, M.H. and M.D. All authors have read and agreed to the published version of the manuscript.

Funding: This study was performed with financial support of Research Program P2-0401 Energy Engineering from Slovenian Research Agency ARRS.

Institutional Review Board Statement: Not applicable.

Informed Consent Statement: Not applicable.

Conflicts of Interest: The authors declare no conflict of interest.

References

1. Platero, C.A.; Sánchez, J.A.; Nicolet, C.; Allenbach, P. Hydropower plants frequency regulation depending on upper reservoir water level. *Energies* **2019**, *12*, 1637. [CrossRef]
2. Yang, W.; Yang, J.; Guo, W.; Norrlund, P. Response time for primary frequency control of hydroelectric generating unit. *Int. J. Electr. Power Energy Syst.* **2016**, *74*, 16–24. [CrossRef]

3. Denholm, P.; O'Connell, M.; Brinkman, G.; Jorgenson, J. *Overgeneration from Solar Energy in California: A Field Guide to the Duck Chart (NREL/TP-6A20-65023)*; National Renewable Energy Laboratory: Golden, CO, USA, 2015.
4. Presas, A.; Valentin, D.; Egusquiza, E.; Valero, C. Detection and analysis of part load and full load instabilities in a real Francis turbine prototype. In *Proceedings of the Journal of Physics: Conference Series*; IOP Publishing: Bristol, UK, 2017; Volume 813.
5. Vagnoni, E.; Favrel, A.; Andolfatto, L.; Avellan, F. Experimental investigation of the sloshing motion of the water free surface in the draft tube of a Francis turbine operating in synchronous condenser mode. *Exp. Fluids* **2018**, *59*, 95. [CrossRef]
6. Podnar, A.; Dular, M.; Širok, B.; Hočevar, M. Experimental Analysis of Cavitation Phenomena on Kaplan Turbine Blades Using Flow Visualization. *J. Fluids Eng.* **2019**, *141*. [CrossRef]
7. Avellan, F. Introduction to cavitation in hydraulic machinery. In *Proceedings of the 6th Int. Conf. on Hydraulic Machinery and Hydrodynamics*; Politehnica University of Timișoara: Timișoara, Romania, 2004; pp. 11–22.
8. Asuaje, M.; Bakir, F.; Kouidri, S.; Rey, R. Inverse Design Method for Centrifugal Impellers and Comparison with Numerical Simulation Tools. In *Proceedings of the International Journal of Computational Fluid Dynamics*; Taylor & Francis: Abingdon-on-Thames, UK, 2004; Volume 18, pp. 101–110.
9. Wu, J.; Shimmei, K.; Tani, K.; Niikura, K.; Sato, J. CFD-based design optimization for hydro turbines. *J. Fluids Eng. Trans. ASME* **2007**, *129*, 159–168. [CrossRef]
10. Zangeneh, M. A compressible three-dimensional design method for radial and mixed flow turbomachinery blades. *Int. J. Numer. Methods Fluids* **1991**, *13*, 599–624. [CrossRef]
11. Zhu, B.; Wang, X.; Tan, L.; Zhou, D.; Zhao, Y.; Cao, S. Optimization design of a reversible pump-turbine runner with high efficiency and stability. *Renew. Energy* **2015**, *81*, 366–376. [CrossRef]
12. Leguizamón, S.; Avellan, F. Open-source implementation and validation of a 3D inverse design method for Francis turbine runners. *Energies* **2020**, *13*, 2020. [CrossRef]
13. Daneshkah, K.; Zangeneh, M. Parametric design of a Francis turbine runner by means of a three-dimensional inverse design method. *IOP Conf. Ser. Earth Environ. Sci.* **2010**, *12*, 012058. [CrossRef]
14. Olimstad, G.; Nielsen, T.; Børresen, B. Dependency on runner geometry for reversible-pump turbine characteristics in turbine mode of operation. *J. Fluids Eng. Trans. ASME* **2012**, *134*. [CrossRef]
15. Xue, P.; Liu, Z.P.; Lu, L.; Tian, Y.J.; Wang, X.; Chen, R. Research and optimization of performances of a pump turbine in pump mode. In *Proceedings of the IOP Conference Series: Earth and Environmental Science*; IOP Publishing: Bristol, UK, 2019; Volume 240.
16. Ma, Z.; Zhu, B.; Rao, C.; Shangguan, Y. Comprehensive hydraulic improvement and parametric analysis of a Francis turbine runner. *Energies* **2019**, *12*, 307. [CrossRef]
17. Lemay, S.; Aeschlimann, V.; Fraser, R.; Ciocan, G.D.; Deschênes, C. Velocity field investigation inside a bulb turbine runner using endoscopic PIV measurements. *Exp. Fluids* **2015**, *56*. [CrossRef]
18. International Electrotechnical Commission. *IEC60193 Hydraulic Turbines, Storage Pumps and Pump-Turbines-Model Acceptance Tests*; BSi Br. Standars; BSI Group: London, UK, 1999.
19. Riegels, F.W. *Aerodynamische Profile*; Oldenbourg: München, Germany, 1958.
20. Althaus, D. *Stuttgarter Profilkatalog 1*; Institut für Aerodynamik und Gasdynamik der Universität Stuttgart: Stuttgart, Germany, 1972.
21. Frikha, S.; Coutier-Delgosha, O.; Astolfi, J.A. Influence of the cavitation model on the simulation of cloud cavitation on 2D foil section. *Int. J. Rotating Mach.* **2008**, *2008*. [CrossRef]
22. Širok, B.; Dular, M.; Stoffel, B. *Kavitacija*; i2: Ljubljana, Slovenia, 2006; ISBN 9616348310. 9789616348317.
23. Chanson, H. Advective diffusion of air bubbles in turbulent water flows. In *Fluid Mechanics of Environmental Interfaces*; Taylor & Francis: Leiden, The Netherlands, 2008; Volume 219, pp. 163–196. ISBN 9781134064236.
24. Deveaux, B.; Fournis, C.; Brion, V.; Marty, J.; Dazin, A. Experimental analysis and modeling of the losses in the tip leakage flow of an isolated, non-rotating blade setup. *Exp. Fluids* **2020**, *61*. [CrossRef]
25. Brennen, C.E. *Cavitation and Bubble Dynamics*; Oxford University Press: New York, NY, USA, 1994.
26. Brennen, C.E. *Fundamentals of Multiphase Flow*; Cambridge University Press: Cambridge, UK, 1995; ISBN 9780511807169.
27. Bajcar, T.; Širok, B.; Eberlinc, M. Quantification of flow kinematics using computer-aided visualization. *Stroj. Vestnik/J. Mech. Eng.* **2009**, *4*, 215–223.

 applied sciences

MDPI

Article

Investigation of the Time Resolution Set Up Used to Compute the Full Load Vortex Rope in a Francis Turbine

Jean Decaix [1,*], Andres Müller [2], Arthur Favrel [3], François Avellan [2] and Cécile Münch-Allig né [4]

1 Institute of Sustainable Energy, School of Engineering, HES-SO Valais-Wallis, Rawil 47, 1950 Sion, Switzerland
2 Laboratory for Hydraulic Machines Ecole Polytechnique Fédérale de Lausanne, Avenue de Cour 33 bis, 1007 Lausanne, Switzerland; andres.mueller@epfl.ch (A.M.); Francois.Avellan@epfl.ch (F.A.)
3 Waseda Research Institute for Science and Engineering, Waseda University, Tokyo 169-8555, Japan; arthur.favrel@aoni.waseda.jp
4 Institute of Sustainable Energy and Institute of Systems Engineering, School of Engineering, HES-SO Valais-Wallis, Rawil 47, 1950 Sion, Switzerland; cecile.muench@hevs.ch
* jean.decaix@hevs.ch

Abstract: The flow in a Francis turbine at full load is characterised by the development of an axial vortex rope in the draft tube. The vortex rope often promotes cavitation if the turbine is operated at a sufficiently low Thoma number. Furthermore, the vortex rope can evolve from a stable to an unstable behaviour. For CFD, such a flow is a challenge since it requires solving an unsteady cavitating flow including rotor/stator interfaces. Usually, the numerical investigations focus on the cavitation model or the turbulence model. In the present works, attention is paid to the strategy used for the time integration. The vortex rope considered is an unstable cavitating one that develops downstream the runner. The vortex rope shows a periodic behaviour characterized by the development of the vortex rope followed by a strong collapse leading to the shedding of bubbles from the runner area. Three unsteady RANS simulations are performed using the ANSYS CFX 17.2 software. The turbulence and cavitation models are, respectively, the SST and Zwart models. Regarding the time integration, a second order backward scheme is used excepted for the transport equation for the liquid volume fraction, for which a first order backward scheme is used. The simulations differ by the time step and the number of internal loops per time step. One simulation is carried out with a time step equal to one degree of revolution per time step and five internal loops. A second simulation used the same time step but 15 internal loops. The third simulations used three internal loops and an adaptive time step computed based on a maximum CFL lower than 2. The results show an influence of the time integration strategy on the cavitation volume time history both in the runner and in the draft tube with a risk of divergence of the solution if a standard set up is used.

Keywords: vortex rope; Francis turbine; cavitation; CFD; RANS

Citation: Lastname, F.; Lastname, F.; Lastname, F. Investigation of the Time Resolution Set Up Used to Compute the Full Load Vortex Rope in a Francis Turbine. *Appl. Sci.* **2021**, *11*, 1168. https://doi.org/10.3390/app11031168

Academic Editor: Florent Ravelet
Received: 2 December 2020
Accepted: 20 January 2021
Published: 27 January 2021

Publisher's Note: MDPI stays neutral with regard to jurisdictional claims in published maps and institutional affiliations.

1. Introduction

The rapidly growing share of new renewable energy sources in the past few decades leads to new challenges for the electrical network stability. Due to their rapid response time, hydraulic power plants are often used to compensate load fluctuations in the grid. However, the injection of the suitable amount of power requires the hydraulic machines to run outside of their Best Efficiency Point (BEP) for which they were designed. As a consequence, the flow leaving the runner possesses a significant residual swirl, leading to an inhomogeneous pressure distribution in the draft tube and eventually to the inception of cavitation.

For Francis turbines, running at an operating point for which the flow rate is above the nominal one leads to the development of an axial vortex rope. If the pressure level inside in the vortex is sufficiently low, cavitation occurs. For a sufficiently high Thoma number, the amount of cavitation is still small and the cavitating vortex rope is stable, whereas, by decreasing further the Thoma number, the vortex rope enters in a self-oscillating mode [1].

On one hand, various studies have been performed to better understand the self-oscillating behaviour of the cavitating vortex rope based on mathematical approaches [2] or 1D analyses [3], experimental investigations [4], 3D CFD [5] and coupling between 1D–3D numerical simulations [6,7]. On the other hand, several investigations look at identifying some key parameters (mainly the mass flow gain factor) useful for the calibration and the development of 1D hydro-acoustic models [8,9] using either experimental measurements [10,11] or CFD simulations [12–14].

Among the 3D CFD studies, the set up for the time integration is not often discussed since it is rare that the mean and the maximum Courant–Friedrichs–Lewy (CFL) numbers or the number of internal loops per time step are provided. However, due to the large range of values that can be taken by the speed of sound in a vapour/liquid mixture and the large time density gradient that can occur in a cell, the time integration requires attention. Usually, in order to improve the accuracy of a simulation at each time step, two ways are available: decreasing the time step or increasing the number of internal loops per time step (if such an option is available). The latter choice can provoke some instabilities [15,16]. In addition, by considering the Ansys CFX software, which is a coupled solver, the equation for the volume fraction is treated in a segregated manner and, in case of the simulation of a rotating turbomachinery, only a first order time scheme is available.

Based on these considerations, this paper investigates the influence of the time resolution set up used to compute an unstable cavitating vortex rope at full load using the Ansys CFX software. The sensitivity to the time integration set up is often disregarded due the large CPU effort required and authors focus rather on grid sensitivity. In addition, sensitivity analysis is often carried out at BEP. In the present paper, the authors focus on an unstable operating point of a Francis turbine at full load, for which a standard set up that allows for accurately capturing the stable cavitating vortex rope [17] is shown to not be able to capture the unstable one. The sensitivity of the time integration set up is performed either by increasing the number of internal loops per time step or by imposing a limit on the time step by limiting the maximum CFL number. The three simulation results are described and some points are discussed by comparisons with experimental data.

2. Test Case

The test case is a 1:16 reduced scale physical model of a Francis turbine with 16 runner blades and 20 guide vanes. The real prototype machine, featuring a rated power of over 444 MW at a specific speed of $\nu = 0.27$ (dimensionless parameters are defined in Equation (1)), experiences severe pressure surges at high load conditions. The load corresponds to 130% of the value at the BEP, i.e., the discharge factor is equal to $Q_{\mathrm{ED}} = 0.26$ instead of 0.20 at the BEP. The speed factor is equal to the rated value of $n_{\mathrm{ED}} = 0.288$. The specific speed of the model runner at these conditions is $\nu_{\mathrm{m}} = 0.31$. For a Thoma number $\sigma = 0.11$, the film displays an unstable cavitating vortex rope (see Figure 1) characterised by a self-excited breathing motion, i.e., a periodic collapse and reformation of the cavity accompanied by large amplitude pressure fluctuations throughout the system. Additional details regarding the experimental set up and the analysis of the measurements are available in [1,4].

The cited dimensionless parameters are defined as follows:

$$\nu = \frac{\omega}{2^{3/4}\,\pi^{1/2}}\frac{Q^{1/2}}{E^{3/4}};\quad Q_{ED} = \frac{Q}{D^2E^{1/2}};\quad n_{ED} = \frac{nD}{E^{1/2}};\quad T_{ED} = \frac{T_m}{\rho D^3 E};\quad \sigma = \frac{NPSE}{E} \tag{1}$$

with ω the rotational speed in rad/s, Q the flow rate in m^3/s, E the total energy in J/kg equal to 262.83, D the outlet runner diameter in m equal to 0.35 m, n the runner frequency in s^{-1} equal to 13.33, and $NPSE$ the Net Positive Suction Energy.

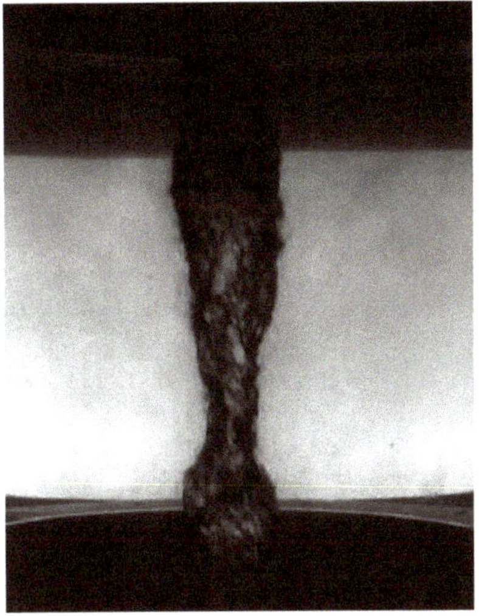

Figure 1. Experimental visualization of the vortex rope at full load.

3. Numerical Settings

The flow is modelled based on the homogeneous Unsteady Reynolds-Averaged Navier–Stokes (U-RANS) equations assuming thermodynamic and mechanical equilibrium [18]. Discarding the energy equation, only the mass and momentum conservation equations are solved. The Reynolds stresses are modelled using the Boussinesq's assumption that introduces a turbulent viscosity μ_t computed with the SST turbulence model [19]. At the wall, the turbulence model is coupled with an automatic wall function. The phase change due to cavitation is modelled using the model proposed by Zwart with the standard values for the parameters [20]. Additional details regarding the flow modelling are available in [17]. Other simulations have also been performed by Wack [21].

The computational domain considered includes a part of the inlet pipe, the spiral case, the stay vanes, the guide vanes, the runner and the draft tube (see Figure 2). The mesh is a structured hexahedral one based on a blocking technique and made with ICEM CFD (see Figure 2). The number of nodes is slightly above 11.5 million with 2.6 million of nodes in the runner domain and 4.3 million in the draft tube. The minimum angle is above 20 degrees and only 9% of the cells have an angle below 50 degrees. The y^+ value does not exceed 85 and the averaged value over each solid walls is lower than 20. A study of the grid sensitivity has been done and published in [17], showing around 2% of differences at the BEP between the computed T_{ED}, Q_{ED} and n_{ED}.

The flow rate is imposed at the inlet, whereas the pressure is set at the outlet. The rotational speed of the runner is imposed on the experimental value. For all the solid walls, a no slip condition is imposed.

The convective fluxes in the momentum equation are computed with a high order scheme [22], whereas the turbulence convective fluxes are computed with an upwind scheme. Regarding the time scheme, the second order backward Euler scheme is applied [23] excepted for the volume fraction equation that is discretized with a first order backward Euler scheme (In Ansys CFX 17.2, for rotating turbomachineries, this option is the only one available.). The time step, the CFL number, the number of internal loops per time step and the time duration of the simulation expressed as the number of runner revolutions achieved are given in Table 1 for each case considered. *Case 1* corresponds to the standard set up in Ansys CFX for a flow in a Francis turbine with a time step set to less than one degree of revolution per time step and five internal loops per time step. This set up has been validated against experimental data for stable cavitating vortex rope in [17]. *Case 2* differs from *Case 1* only by the number of internal loops, which is set to 15. *Case 3* differs from the other two cases since the time step is not imposed by the user but is computed automatically by imposing a limitation on the CFL number to a value of 2. This value is sufficiently small to deal with the accuracy of the first order backward Euler scheme used for the volume fraction equation keeping advantage of an implicit formulation. In addition, the number of internal loops is decreased to a value of 3 even if two internal loops are enough to reach a second order time accuracy (see [24]) for the equations discretized with a second order scheme. The unsteady simulations are started from a converged steady simulation (see [17]).

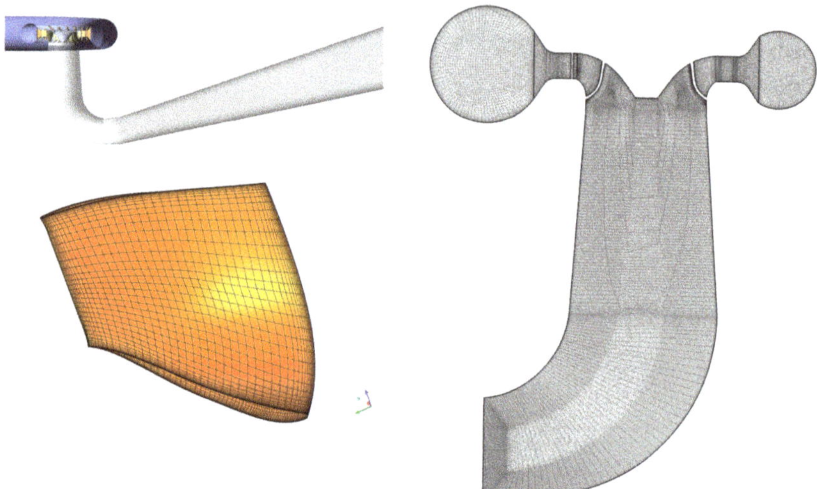

Figure 2. Computational domain (left top), view of the surface mesh of blade (left bottom) and side view of the structured mesh in a mid-plane (right).

Even if CFX used a coupled solver, the volume fraction equation is treated in a segregated manner. The interfaces between the stationary and rotating domains are handled using a General Grid Interface (GGI) algorithm.

Table 1. Parameters set for the time integration. Bold terms refer to the imposed values by the user.

Simulation	Time Step (deg/Δt)	CFL (-)	Internal Loop (-)	Number of Revolution (-)	CPU Effort (Hours)
Case 1	**0.96**	80	**5**	2.3	3′897
Case 2	**0.96**	80	**15**	13.3	76′861
Case 3	adaptive	$\leq$ **2**	3	3.15	137′444

4. Results

First of all, it has to be mentioned that the simulation *Case 1* diverges when the vortex rope in the draft tube collapses (see the next section for a detailed discussion). However, until this point, all the simulations show the development of a cavitating vortex rope and a cavitation sheet over the trailing edge of the suction side (see Figure 3). Such features are supported by experimental visualisation [4]. Compared to the experiment (Figure 1), the volume of vapour inside the vortex rope is underestimated.

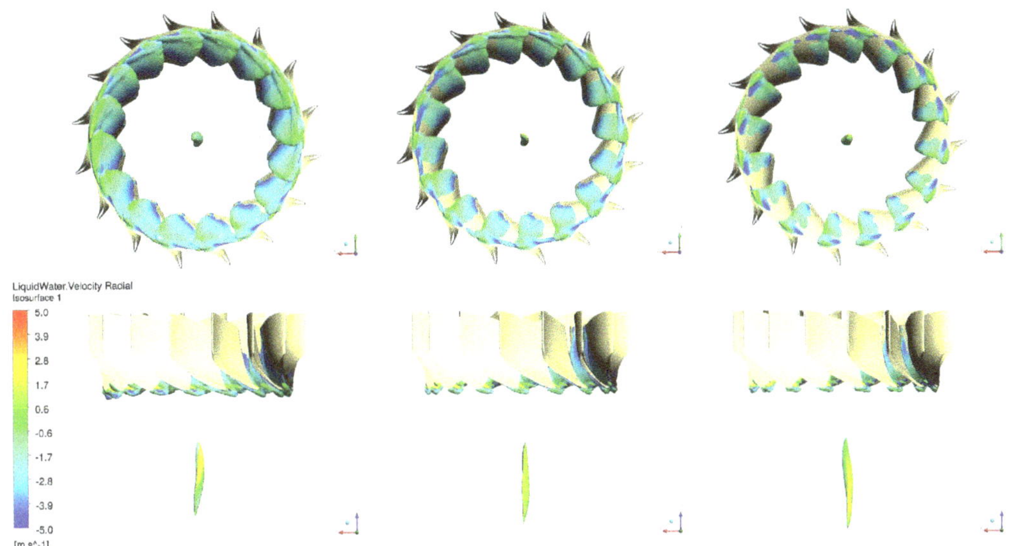

Figure 3. Instantaneous iso-surface of the liquid volume fraction $\alpha_L = 0.9$. Downstream view (top) and side view (bottom).

4.1. Integrated Variables

The time-history of the efficiency, the dimensionless volume of vapour in the computational domain, the torque factor T_{ED} and the speed factor n_{ED} are plotted on Figure 4 (the experimental values for the T_{ED} and n_{ED} are added in the caption of the figure). The divergence of the simulation *Case 1* just after the collapse of the cavitation volume is clearly put in evidence. On the contrary, simulations *Case 2* and *Case 3* do not show a complete collapse of the cavitation volume even if a time oscillation of the different quantities is observed. The first cycle seems to be shortened by increasing the requirement on the time integration since it is shorter for *Case 2* compared to *Case 1* and the shortest for *Case 3*.

Regarding the efficiency, the torque factor and the speed factor, the differences between the simulations are around 1.5%, whereas, for the volume of vapour, simulation *Case 1* predicts a volume that is at least four times higher. Compared to the measurements, the T_{ED} and n_{ED} are overestimated respectively by approximately 2.5% and 3.5%. Simulations *Case 2* and *Case 3* have a rather similar behaviour even if a time shift is observed due to a faster first cycle for *Case 3*. Furthermore, due to the smaller time step imposed in *Case 3*, fluctuations at higher frequencies are observed in all the signals.

For simulation *Case 2* between the seventh and eleventh runner revolution, the slow variation of the efficiency, the torque factor, and the speed factors disappear, whereas a fluctuation at a low frequency is still observed for the volume of vapour. By splitting the time history of the volume of vapour between the volume in the runner and draft tube domains (see Figure 5), it is noticeable that between the seventh and eleventh runner revolution, the volume of vapour in the draft tube is close to zero and then increases again after the tenth runner revolution. Therefore, the low frequency seems to be related to the volume of vapour in the draft tube, which means with the presence of a cavitating vortex rope. This point seems to be in agreement with the experimental investigations [1].

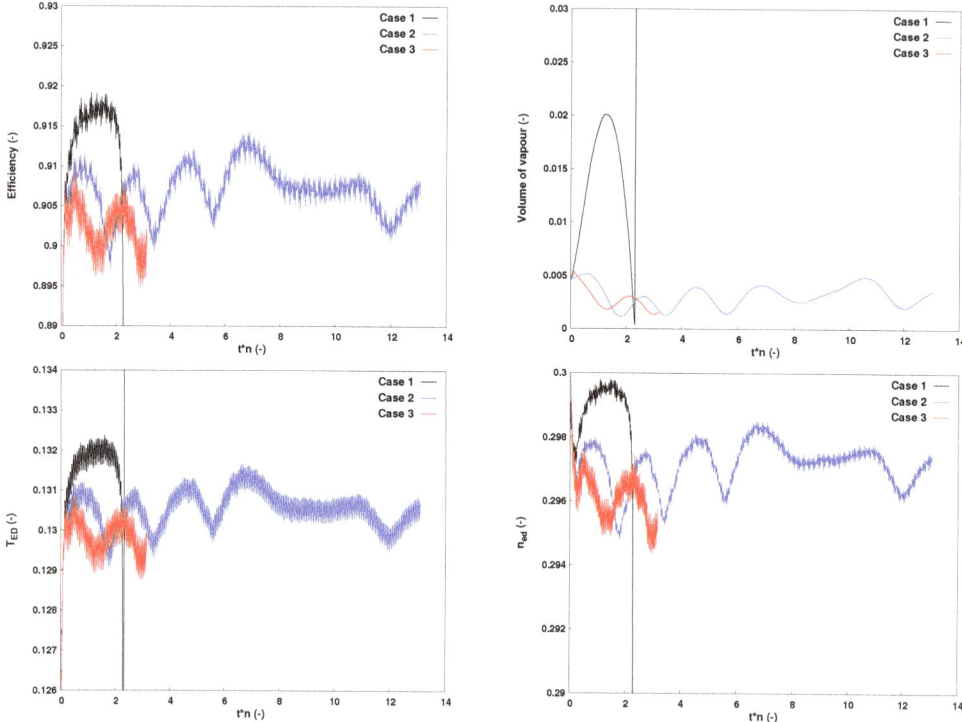

Figure 4. Time-history of the efficiency, the dimensionless volume of vapour, the torque factor T_{ED} (measured value: 0.127) and the speed factor n_{ED} (measured value: 0.288).

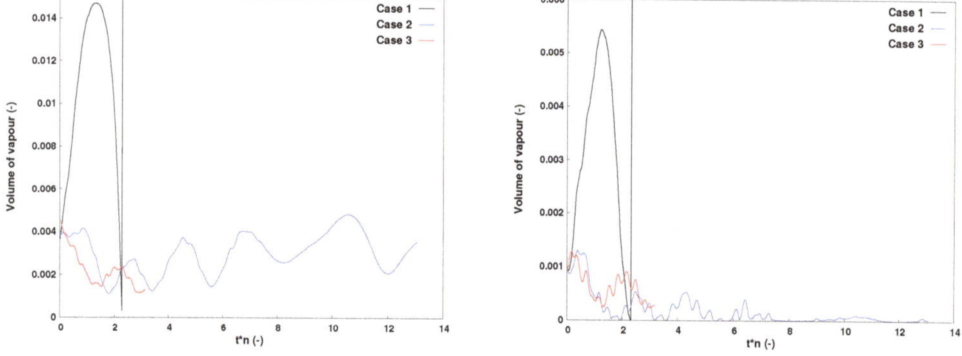

Figure 5. Time-history of the dimensionless volume of vapour in the runner (left) and in the draft tube (right).

A Fourier analysis of the efficiency signal is shown in Figure 6 on the left. The frequency is divided by the runner frequency f_n. The modes put in evidence are the ones associated with the blade passage frequency $f_b = 16 f_n$ and the first and fifth harmonic. Both *Case 2* and *Case 3* capture the fifth harmonic, which corresponds to the rotor/stator interaction since $5\,f_b = 4\,f_{gv} = 80 f_n$. A Fourier analysis of the volume of vapour has also been carried out even if only the simulation *Case 2* has achieved enough cycles to provide an estimation of the low frequency characteristic of the time evolution of the volume of vapour. By focusing only on *Case 2*, a low frequency around $0.3 f_n$ is put in evidence, which is closed to $0.25 f_n$, the value observed on the torque during the measurements [1].

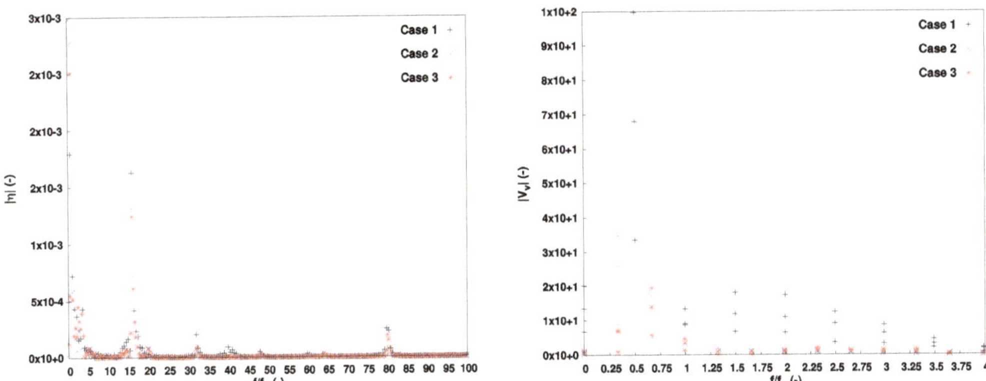

Figure 6. Magnitude of the Fourier transform of the efficiency (left) and volume of vapour (right) signals.

4.2. Local Pressure

Several pressure probes have been set in the computational domain as shown in Figure 7. Two probes are located in two different guide vane channels, six are located on a runner blade with three on the pressure side and three others on the suction side, two times four probes are distributed at ninety degrees on the wall in two sections of the draft tube cone (as in the experiment) and finally two probes are set in the draft tube elbow.

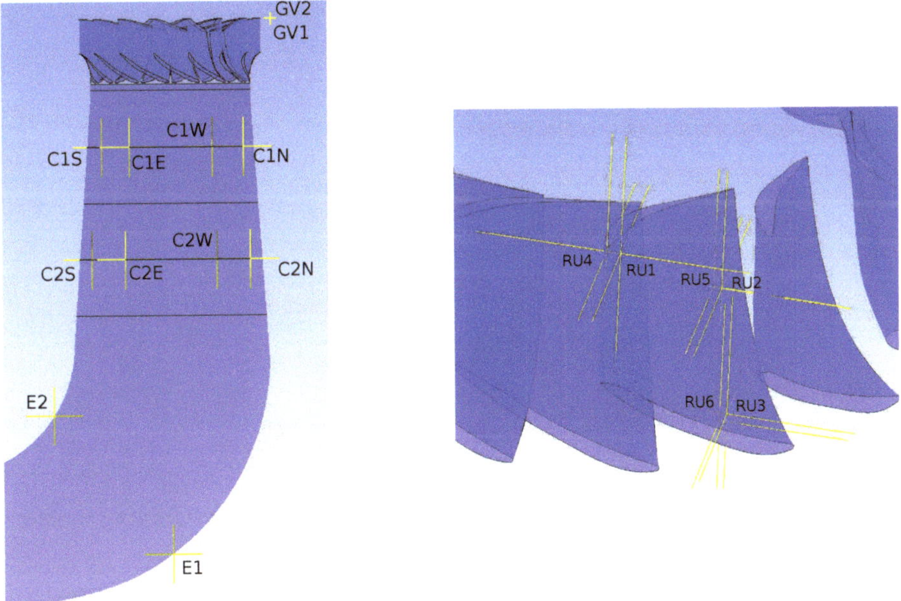

Figure 7. Positions of pressure probe in the computational domain (left). Zoom on the runner probe locations (right).

For each probe, the fluctuations of the pressure coefficient are computed by subtracting the average pressure at the probe location and then by dividing the result by $0.5\,\rho\,U_{\overline{1}}^2$ (with $U_{\overline{1}}$ the peripheral velocity at the runner outlet). In addition, for the probes in the cone of the draft tube, a spatial averaging is also performed.

The pressure signals on the pressure side of the runner at the leading edge (probe Ru1) and on the suction side of the trailing edge (probe Ru6) and in the two sections of the draft tube cone are displayed in Figure 8. In the draft tube cone, the experimental pressure signals are also added.

At the leading edge of the runner, the pressure fluctuations captured by the three simulations are identical with a high frequency content. At the trailing edge, the fluctuations at high frequency alternate with a flat period corresponding to the period during which the probe is covered by a pure vapour sheet. No clear pattern can be observed. In the draft tube cone, compared to the experiment, the simulations *Case 2* and *Case 3* are smoother without a strong peak of pressure after the partial collapse of the vortex rope. Simulation *Case 1* presents a strong pressure peak at the collapse of the vortex rope, which seems to follow qualitatively the trends of the experimental one. However, as already mentioned, the simulation diverges after the collapse.

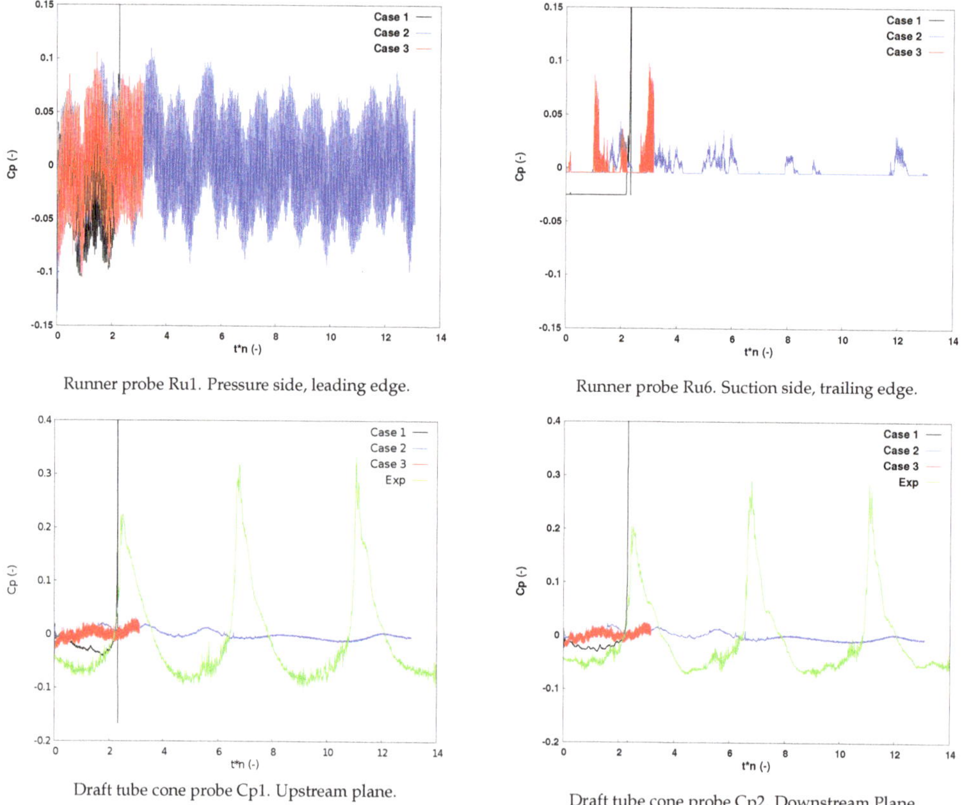

Figure 8. Time history of the pressure fluctuations in the runner and the draft tube cone. The zero of experimental signal has been set arbitrary.

The Fourier transforms of the pressure signals in the runner and the draft tube are represented in Figure 9. At the leading edge of the runner, the guide vane passage frequency $f_{gv} = 20f_n$ and its harmonic until the fourth are clearly observed. At the trailing edge, only the fundamental is observed. In addition, a frequency around $120f_n$ is captured by the simulations *Case 2* and *Case 3*. This frequency is also observed in the cone of the draft tube but with an attenuation. The origin of this frequency is not clear.

In the draft tube, the comparison of the simulation *Case 2* with the experiment makes the presence of the low frequency appear around $0.3f_n$ in the simulation instead of $0.25f_n$ in the experiment. Obviously, the amplitude is low compared to the experiment due to the inability of the simulation to capture accurately the collapse of the vortex rope as shown in Figure 8.

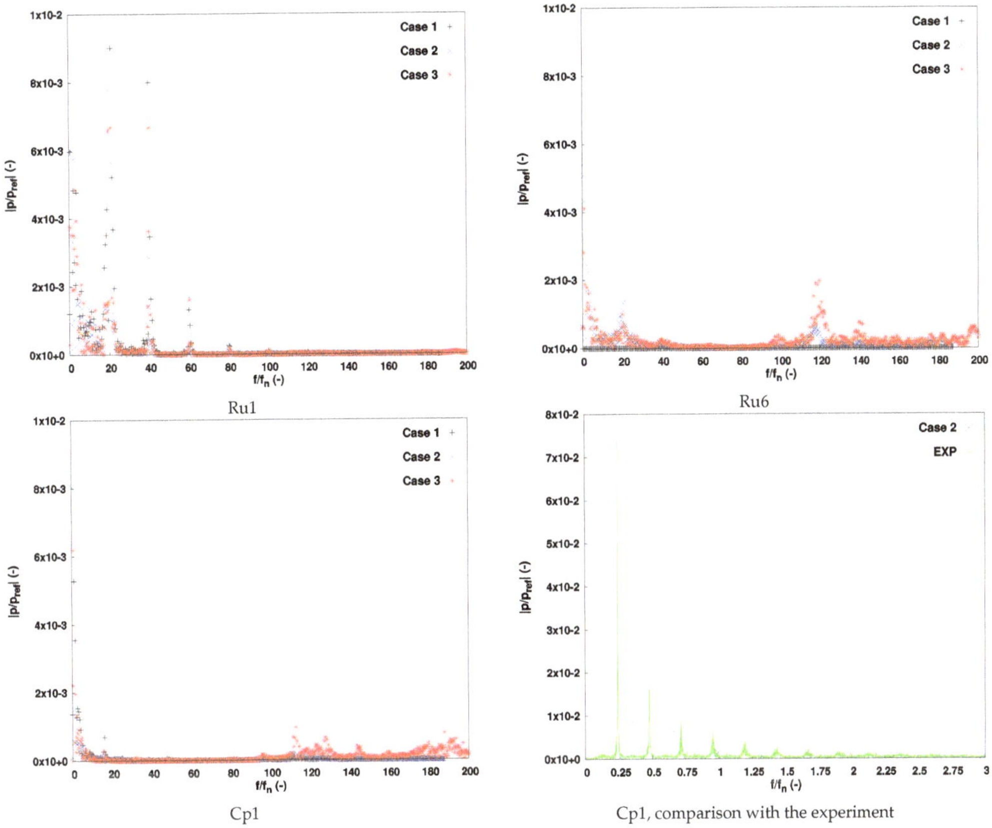

Ru1

Ru6

Cp1

Cp1, comparison with the experiment

Figure 9. Magnitude of the Fourier transform of the pressure signal in the runner and the draft tube.

5. Discussion

The results described in the previous section make a question appear—why the simulation *Case 1* seems to follow more closely the behaviour of the vortex rope observed experimentally than the simulations *Case* 2 and *Case* 3, which use a refined time integration set up.

The Root Mean Square (RMS) and the maximum RMS of the pressure equation are compared in Figure 10. The RMS reach values lower than 10^{-3} for all the simulations except simulation *Case 1* before the simulation crashes. For *Case* 2, the residuals reach values below 10^{-5}, which shows the role plays by the number of internal loops per time step. The maximum RMS are quite high for all simulations with some values above 0.1. Interestingly, the maximum RMS of *Case 1* and *Case* 3 are around 0.4, but *Case* 3 does not crash.

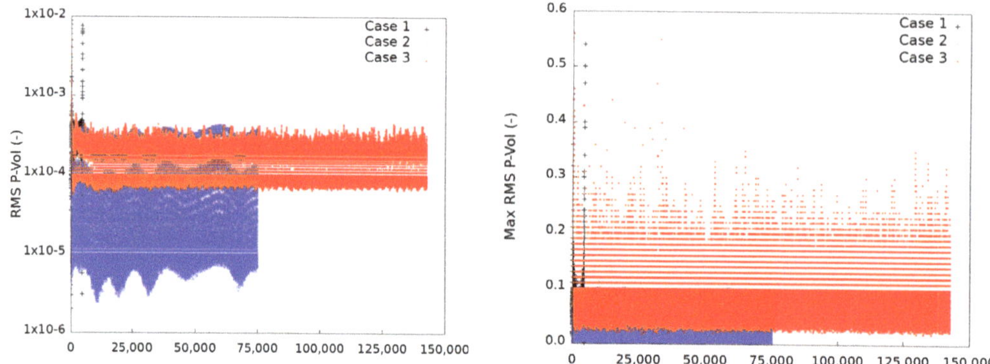

Figure 10. RMS (left) and maximum RMS (right) of the pressure equation.

The time history of the dimensionless mass flow $Q^* = Q/Q_{inlet}$ at the inlet of the runner and at the outlet of the computational domain are compared for the three simulations in Figure 11. In the runner inlet section, upstream from the region where cavitation occurred, the mass flow is constant and equal to 1. Downstream, due to the phase change, the mass flow oscillates. The oscillation is stronger for *Case 1* with a strong decrease of the mass flow, which is a consequence of the vapour volume growth (see Figure 5).

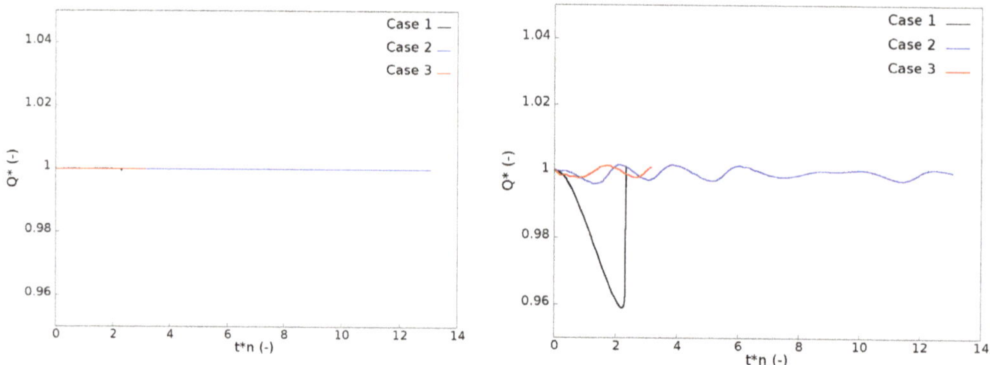

Figure 11. Time history of the dimensionless mass flow at the runner inlet section (left) and at the outlet of the computational domain (right).

By assuming a density of the vapour negligible compared to the liquid density, the 1D mass conservation equation applied for the whole computational domain writes:

$$Q_1 - Q_2 = -\frac{dV}{dt} \tag{2}$$

with Q_1 and Q_2, respectively, the mass flow rate at the inlet and outlet of the computational domain and V the volume of vapour. The left- and right-hand terms of Equation (2) are plotted in Figure 12 for the three cases. For *Case 2* and *3*, the mass conservation is respected contrary to *Case 1* for which the time variation of the volume of vapour is smaller than the difference of the flow rate. This mass unbalanced seems to be responsible for the collapse of the vortex rope and the crash of the simulation. As the boundary condition are fixed, the mass flow rate unbalanced cannot be supported infinitely since it is a non-equilibrium state. Therefore, the collapse of the vortex is the only way to return to an equilibrium state.

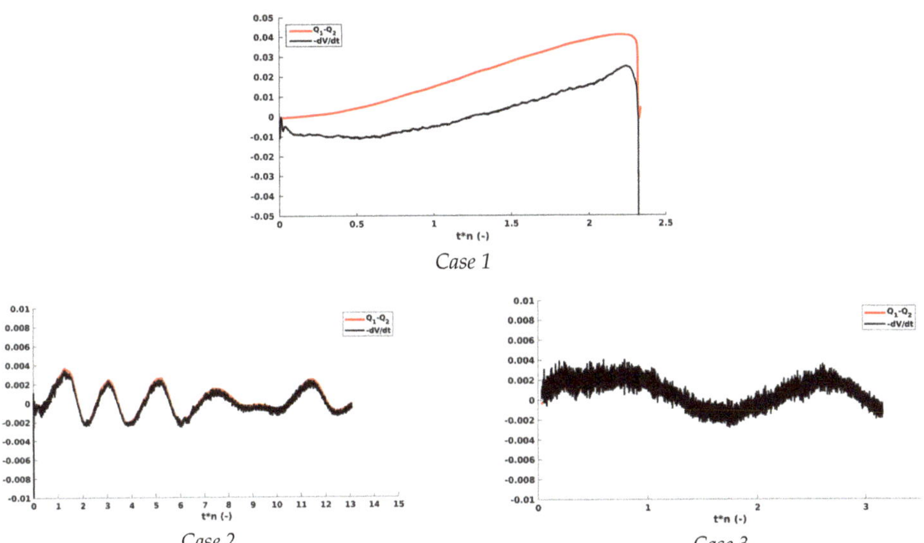

Figure 12. Time history of the left- and right- hand terms of Equation (2) for each case.

6. Conclusions

The influence of the time resolution set up on the simulation of an unstable cavitating rope has been performed using the Ansys CFX software. A standard set up *Case 1* usually used for the URANS computation of a Francis turbine is not able to capture the flow due to the crash of the simulation after the first collapse of the cavitating vortex rope. By increasing the number of internal loops from 5 to 15, the simulation *Case 2* is able to capture several cycles of the vortex rope with a dampened amplitude compared to the experiment but with a frequency close to the measured one. It has also been noticed that the low frequency observed on the torque or the efficiency seems to be related to the presence of a cavitating vortex rope in the draft tube since, in its absence, the low frequency disappears. The simulation *Case 3* carried out with an imposed maximum CFL number lower than 2, i.e., with a smaller time step, and only three internal loops does not provide a dynamic behaviour very different than *Case 2* even if a longer simulation could show differences.

Surprisingly, the simulation *Case 1* is the only one that gives a duration and an amplitude of the pressure peak in agreement with the experiment. Nevertheless, the simulation stops when the vortex rope collapse due to a solver crashes and no successive cycles are computed. On the contrary, the simulations *Case 2* and *Case 3* do not show any growth or collapse of the cavitating vortex rope. This difference is explained by the fact that the simulation *Case 1* does not respect the mass conservation equation at each time step contrary to the two other simulations.

To compute an unstable vortex rope, it is therefore necessary to take care of the time integration set up by increasing at least the number of internal loops. To improve the accuracy of the simulation, the U-RANS simulation needs, in addition, to be coupled with a 1D hydroacoustic model allowing for adjusting the boundary conditions at the inlet and outlet boundaries of the computational domain [6,7].

Author Contributions: J.D. contributed to carrying out the simulations and the post-treatment of the CFD results. He is the main writer of the paper. A.M. and A.F. performed the experimental work and provide the experimental results used in this paper. They also contribute to the redaction of the paper. F.A. and C.M.-A. were responsible for the supervision, project administration, and the funding acquisition. All authors have read and agreed to the published version of the manuscript.

Funding: The research leading to the results published in this paper is part of the HYPERBOLE research project, granted by the European Commission (ERC/FP7-ENERGY-2013-1-Grant 608532)

Institutional Review Board Statement: Not applicable.

Informed Consent Statement: Not applicable.

Data Availability Statement: The data presented in this study could be available on request from the corresponding author. The data are not publicly available due to NDA.

Conflicts of Interest: The authors declare no conflict of interest. The funders had no role in the design of the study; in the collection, analyses, or interpretation of data; in the writing of the manuscript, or in the decision to publish the results.

References

1. Müller, A.; Favrel, A.; Landry, C.; Avellan, F. Fluid-structure interaction mechanisms leading to dangerous power swings in Francis turbines at full load. *J. Fluids Struct.* **2017**, *69*, 56–71. [CrossRef]
2. Susan-Resiga, R.; Muntean, S.; Avellan, F.; Anton, I. Mathematical modelling of swirling flow in hydraulic turbines for the full operating range. *Appl. Math. Model.* **2011**, *35*, 4759–4773. [CrossRef]
3. Chen, C.; Nicolet, C.; Yonezawa, K.; Farhat, M.; Avellan, F.; Tsujimoto, Y. One-Dimensional Analysis of Full Load Draft Tube Surge. *ASME J. Fluids Eng.* **2008**, *130*, 041106. [CrossRef]
4. Müller, A.; Favrel, A.; Landry, C.; Yamamoto, K.; Avellan, F. On the physical mechanisms governing self-excited pressure surge in Francis turbines. *IOP Conf. Ser. Earth Environ. Sci.* **2014**, *22*, 032034. [CrossRef]
5. Tamura, Y.; Tani, K. Investigation on internal flow of draft tube at overload condition in low specific speed Francis turbine. *IOP Conf. Ser. Earth Environ. Sci.* **2016**, *49*. [CrossRef]
6. Chirkov, D.; Avdyushenko, A.; Panov, L.; Bannikov, D.; Cherny, S.; Skorospelov, V.; Pylev, I. CFD simulation of pressure and discharge surge in Francis turbine at off-design conditions. *Iop Conf. Ser. Earth Environ. Sci.* **2012**, *15*. [CrossRef]
7. Mössinger, P.; Conrad, P.; Jung, A. Transient two-phase CFD simulation of overload pressure pulsation in a prototype sized Francis turbine considering the waterway dynamics. *IOP Conf. Ser. Earth Environ. Sci.* **2014**, *22*, 032033. [CrossRef]
8. Koutnik, J.; Nicolet, C.; Schohl, G.; Avellan, F. Overload Surge Event in a Pumped-Storage Power Plant. In Proceedings of the 23rd IAHR Symposium on Hydraulic Machinery and Systems, Ljubljana, Slovenia, 17–21 October 2006; Volume 1, pp. 1–15.
9. Alligné, S.; Nicolet, C.; Tsujimoto, Y.; Avellan, F. Cavitation Surge Modelling in Francis Turbine Draft Tube. *J. Hydraul. Res.* **2014**, *52*, 1–13. [CrossRef]
10. Müller, A.; Alligné, S.; Paraz, F.; Landry, C.; Avellan, F. Determination of Hydroacoustic Draft Tube Parameters by High Speed Visualization during Model Testing of a Francis Turbine. In Proceedings of the 4th International Meeting on Cavitation and Dynamic Problems in Hydraulic Machinery and Systems, Belgrade, Serbia, 26–28 October 2011.
11. Müller, A.; Yamamoto, K.; Alligné, S.; Yonezawa, K.; Tsujimoto, Y.; Avellan, F. Measurement of the Self-Oscillating Vortex Rope Dynamics for Hydroacoustic Stability Analysis. *ASME J. Fluids Eng.* **2016**, *138*, 021206. [CrossRef]
12. Flemming, F.; Foust, J.; Koutnik, J.; Fisher, R.K. Overload Surge Investigation Using CFD Data. *Int. J. Fluid Mach. Syst.* **2009**, *2*, 315–323. [CrossRef]
13. Dörfler, P.; Keller, M.; Braun, O. Francis full-load surge mechanism identified by unsteady 2-phase CFD. *IOP Conf. Ser. Earth Environ. Sci.* **2010**, *12*, 012026. [CrossRef]
14. Alligné, S.; Decaix, J.; Müller, A.; Nicolet, C.; Avellan, F.; Münch, C. RANS computations for identification of 1D cavitation model parameters: Application to full load cavitation vortex rope. *IOP Conf. Ser. Earth Environ. Sci.* **2016**, *49*, 082014. [CrossRef]
15. Pulliam, T.H. Time accuracy and the use of implicit methods. In Proceedings of the 11th Computational Fluid Dynamics Conference, Orlando, FL, USA, 6–9 July 1993; pp. 685–693. [CrossRef]
16. Chiew, J.J.; Pulliam, T.H. Stability Analysis of Dual-time Stepping. In Proceedings of the 46th AIAA Fluid Dynamics Conference, Washington, DC, USA, 13–17 June 2016; [CrossRef]
17. Decaix, J.; Müller, A.; Favrel, A.; Avellan, F.; Münch, C. URANS Models for the Simulation of Full Load Pressure Surge in Francis Turbines Validated by Particle Image Velocimetry. *J. Fluids Eng.* **2017**, *139*. [CrossRef]
18. Brennen, C. *Cavitation and Bubble Dynamics*; Oxford University Press: New York, NY, USA, 1995.
19. Menter, F. Zonal Two Equation k-ω turbulence models for aerodynamic flows. In Proceedings of the 24th Fluid Dynamics Conference, Orlando, FL, USA, 6–9 July 1993.
20. Zwart, P.; Gerber, A.; Belamri, T. A two-phase flow model for predicting cavitation dynamics. In Proceedings of the Fifth International Conference on Multiphase Flow, Yokohama, Japan, 30 May–3 June 2004; Volume 152.

21. Wack, J.; Riedelbauch, S. Two-phase simulations of the full load surge in Francis turbines. *IOP Conf. Ser. Earth Environ. Sci.* **2016**, *49*. [CrossRef]
22. Barth, T.J.; Jesperon, D.C. The Design and Application of Upwind Schemes on Unstructured Meshes. In Proceedings of the 27th Aerospace Sciences Meeting, Reno, NV, USA, 9–12 January 1989.
23. Hirsch, C. *Numerical Computation of Internal and External Flows—Fundamentals of Numerical Discretization*; John Wiley & Sons: Hoboken, NJ, USA, 1988; Volume I.
24. Menter, F. CFD Best Practice Guidelines for CFD Code Validation for Reactor-Safety Applications. In 5th *Framework Programme of the European Atomic Energy Community (Euratom) for Research and Training Activities, 1998–2002*; Technical Report; European Commission: Brussels, Belgium, 2002.

Article

Numerical Investigation of Methodologies for Cavitation Suppression Inside Globe Valves

Jun-ye Li [1], Zhi-xin Gao [1,2,*], Hui Wu [1] and Zhi-jiang Jin [2]

1 SUFA Technology Industry Co., Ltd., CNNC, Suzhou 215129, China; lijy@chinasufa.com (J.-y.L.); wuh@chinasufa.com (H.W.)

2 Institute of Process Equipment, College of Energy Engineering, Zhejiang University, Hangzhou 310027, China; jzj@zju.edu.cn

* Correspondence: zhixingao@zju.edu.cn

Received: 20 July 2020; Accepted: 7 August 2020; Published: 11 August 2020

Abstract: Cavitation inside globe valves, which is a common phenomenon if there is a high-pressure drop, is numerically investigated in this study. Firstly, the cavitation phenomenon in globe valves with a different number of cages is compared. When there is no valve cage, cavitation mainly appears at the valve seat, the bottom of the valve core, and the downstream pipelines. By installing a valve cage, cavitation bubbles can be restricted around the valve cage protecting the valve body from being damaged. Secondly, the effects of the outlet pressure, the working temperature, and the installation angle of two valve cages in a two-cage globe valve are studied to find out the best method to suppress cavitation, and cavitation number is utilized to evaluate cavitation intensity. Results show that cavitation intensity inside globe valves can be reduced by increasing the valve outlet pressure, decreasing the working temperature, or increasing the installation angle. Results suggest that increasing the outlet pressure is the most efficient way to suppress cavitation intensity in a globe valve, and the working temperature has a minimal effect on cavitation intensity.

Keywords: cavitation; cavitation number; globe valve; valve cage; computational fluid dynamics

---

1. Introduction

For the valves applied in the industrial pipelines, there is often a reduction of the flow channel around the valve seat or the valve plug, where the liquid pressure may fall below the saturated vapor pressure and the cavitation phenomenon may occur. Hence cavitation is one of the reasons that can lead to valve failure if a valve is not designed appropriately [1].

Valves play important roles in many working scenarios, and common applications include flow, pressure, and temperature control; common valves including ball valves [2], control valves [3], and globe valves [4], etc. are summarized and shown in Table 1. Works about valves mainly focus on the flow characteristics and structure optimization. Qian et al. [5] investigated the acoustic power with different thermal conditions in a temperature and pressure regulation valve. Fratini and Pasta [6] investigated the effects of the residual stress on the fatigue crack. Wang et al. [7] designed a new spool structure and various pressure and valve openings were investigated to find out their effects on the stability of the flow field. Lin et al. [8] studied the two-phase flow characteristics of a gate valve and the effects of size and numbers of the particles on erosion were analyzed. Jin et al. [9] and Qian et al. [10–12] applied the Tesla valves in the hydrogen decompression process and the flow characteristics and the aerodynamic noise of a single and a multi-stage Tesla valve was investigated. Fan et al. [13] focused on the contaminated friction of a spool valve. Qian et al. [3] investigated the flow characteristics of a feed-water valve with different structures of the throttling window. Yan et al. [14] focused on the erosion in a servo valve. Pasta et al. [15] and Rinaudo et al. [16] investigated the hemodynamics in biomedical heart valves and the results could be helpful for the clinical diagnosis. Zhang et al. [17] studied the effects of the damping sleeve on the flow force of an on-off valve. Zhong et al. [18] proposed a two-level fuzzy control algorithm based on the calculated flow rate feedback from the spool displacement to realize high precision flow control.

Table 1. Common valves applied in different industries.

Valve Type	Ball Valve [2] [1]	Control Valve	Globe Valve
Structure			

[1] Reproduced from reference [2]. Copyright 2017, Elsevier.

Cavitation that appears in the valve can lead to self-excited noise, cavitation erosion, and system jam, etc. [19,20], and it may even cause severe damage to the valve spool as shown in Figure 1 [21]. To date, more and more works focusing on the cavitation problem in various kinds of valves can be found, and an experimentally verified computational fluid dynamics (CFD) method is becoming a popular tool [22,23]. Wen et al. [24] studied the velocity of a needle valve motion on cavitation using the dynamic grid overlapping techniques and a visual experimental platform. Li et al. [25] focused on the cavitation in bileaflet mechanical heart valves and their results showed that it was induced by the high closing velocity and water hammer effect. Jin et al. [26,27] investigated the cavitation in a globe valve and a sleeve regulating valve, and the effects of the valve core shape and the inlet velocity were obtained. Liu et al. [28] studied the effects of inlet pressure on cavitation location and area in a regulating valve, and Qiu et al. [29] focused the effects of valve opening and pressure drop of a sleeve regulating valve. Liu et al. [30] investigated the cavitation erosion in a butterfly valve with various inlet pressure and valve opening angles. Jin et al. [31] numerically analyzed the cavitating flow in microchannels and the effects of the diameter ratio and pressure drop were discussed. Zhang [32] studied the flow characteristics of a cone throttle valve under cavitation, and the cavitation luminescence was used to help relative experiments. Yuan et al. [33] compared the compressible and incompressible method, and their results showed that the compressible method could provide more precise results.

Figure 1. Damage of the valve plug results from cavitation erosion [21]-Copyright 2016, Elsevier.

To suppress or prevent the occurrence of cavitation, valve structure optimization, adding outlet pressure, or adding stage/anti-cavitation trims is inevitable. Yaghoubi et al. [4] numerically studied the effects of the numbers of valve trims on cavitation intensity, and their results showed that the numbers of valve trims should be below two. Xu et al. [34] performed structure optimization of the valve port in a relief valve to reduce cavitation intensity. Shi et al. [35] utilized a drainage device to suppress the cavitation in a throttle valve, and cavitation intensity could be reduced with appropriate inlet and outlet pressure. Lee et al. [36] optimized the structure of the bottom plug in a three-way reversing valve to suppress valve cavitation. Chern et al. [37] found that by adding a valve cage, cavitation could be limited to the region close to the valve cage, and the valve body could be protected from damage resulting from valve cavitation.

Globe valves are widely used in nuclear and thermal power stations, or in petrochemical fields, where there is often a high-pressure drop condition resulting in cavitation inside globe valves [37]. In this paper, a globe valve with two valve cages is proposed to suppress valve cavitation, and each cage has 20 evenly distributed slender orifices. Firstly, the flow characteristics of the globe valve with and without valve cage are compared to investigate cavitation distribution and cavitation intensity. Then the effects of the working temperature and the valve outlet pressure on cavitation intensity for the globe valve with two valve cages are studied. At last, the effects of the installation angle of two valve cages are analyzed and the most efficient method to reduce cavitation intensity is found out.

2. Numerical Methods

The structure of the investigated globe valve is shown in Figure 2; here the globe valve with a different number of cages and the geometry of the cage are displayed. The two-stage globe valve is self-designed to be used in the pipe system with a high-pressure drop, and the model is created using commercial software SOLIDWORKS. The nominal pipe diameter of the globe valve, D, is 50 mm, and the entrance pipe length is 3 D and the downstream pipe length is 8 D. The throttling orifices on each cage are the same and are evenly distributed along the circumference. The throttling orifices on two cages can either be installed aligned or be installed with an installation angle, α, as shown in Figure 2.

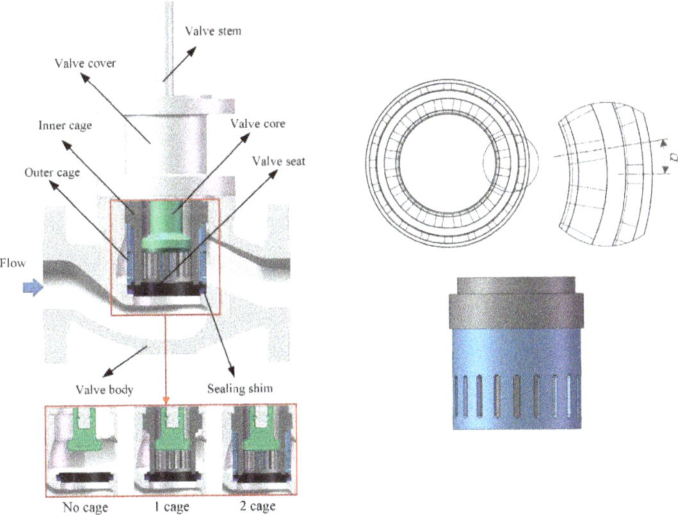

Figure 2. Structure of the investigated globe valve.

Fluid flow through the two-cage globe valve with a fully opened state is numerically solved. Liquid water and water vapor under different temperatures are considered as the incompressible working fluid, and the properties of the working fluid are shown in Table 2. The continuity and the momentum equations are solved using commercial software ANSYS Fluent, and the governing equations and the equations of the Realizable k-ε turbulence model can be found in the previous work [26]. The cavitation model proposed by Zwart et al. [38] is used and the equations describing cavitation are shown below.

$$\frac{\partial}{\partial t}(\alpha\rho_v) + \nabla(\alpha\rho_v u_v) = R_e - R_c \tag{1}$$

$$R_e = F_{vap} \frac{3\alpha_{nuc}(1-\alpha_v)\rho_v}{R_B} \sqrt{\frac{2}{3}\frac{p_v - p}{\rho_l}} \quad (p \le p_v) \tag{2}$$

$$R_c = F_{cond} \frac{3\alpha_v \rho_v}{R_B} \sqrt{\frac{2}{3}\frac{p - p_v}{\rho_l}} \quad (p \ge p_v) \tag{3}$$

here α is vapor volume fraction, subscript v stands for vapor phase, ρ stands for density, u stands for velocity, R_e and R_c are related to the growth and collapse of the vapor bubbles, R_B, F_{vap}, and F_{cond} are model constant. During the simulation, the pressure-based SIMPLE (Semi-Implicit Method for Pressure Linked Equations) algorithm is utilized, and the second-order upwind discretization scheme is applied. The residuals are set as 0.001 and the area-weighted inlet velocity is monitored, and the solution is considered as converged if the residual is below 0.001 and the inlet velocity is constant.

Table 2. Properties of water and water vapor.

		20 °C	40 °C	60 °C	80 °C
Water	Density/kg·m^{-3}	998.16	992.18	983.17	971.78
	Viscosity/pa·s	0.0010016	0.0006529	0.000466	0.000354
Water vapor	Density/kg·m^{-3}	0.0173	0.05124	0.1304	0.294
	Viscosity/pa·s	9.78×10^{-6}	1.03×10^{-5}	1.09×10^{-5}	1.16×10^{-5}
Saturated pressure/Pa		2339	7384	19945.8	47414.7

Pressure inlet and pressure outlet are applied as the main boundary conditions. The inlet pressure is 1.17 MPa, and the outlet pressure varies from 0.1 to 0.25 MPa. The no-slip wall with the scalable wall function is used. The temperature variation is neglected during the simulation.

A polyhedral mesh with boundary layer is utilized in this study, and the zone near the throttling window of the cage is refined. Figure 3 shows the mesh of the investigated two-cage globe valve. Mesh independence is done with a different number of grid cells when the outlet pressure is 0.25 MPa and the temperature is 80 °C. The inlet velocity and pressure drop are compared and shown in Table 3, and it is clear that the results are close when the cell number is 824,206 and 1,550,930, thus 824,206 cells are adopted eventually.

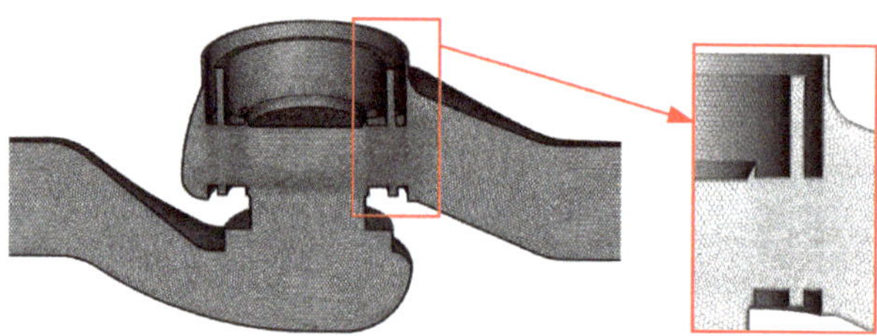

Figure 3. Mesh of the two-cage globe valve.

Table 3. Grid independence check.

Cells Number	Pressure Drop/Pa	Error/%	Velocity/m·s^{-1}	Error/%
453,437	884,382.9	1.15%	16.48	−3.5%
824,206	874,309.2	-	17.08	-
1,550,930	868,852.6	−0.62%	17.39	1.8%

3. Results and Discussion

The globe valve with and without cages is compared in this study, and the installation angle between two cages, the outlet pressure, and the working temperature are investigated. Flow on the middle plane of the valve body including the velocity and pressure distributions and the streamlines are analyzed to help understanding the cavitation phenomenon. Cavitation number, loss coefficient, and induced water vapor volume are quantitatively studied and compared.

3.1. Comparison of Cavitation in Globe Valves with and without Cages

When there is a large pressure drop, local pressure is very likely to drop under the saturated vapor pressure. Here, the globe valve without a cage, the globe valve with one cage and the globe valve with two cages are simply compared to find out the effects of the cage. The outlet pressure is chosen as 0.1 MPa and the working temperature is 80 °C.

The velocity and pressure distributions in the globe valve with a different number of cages are shown in Figure 4, and an obvious difference in the velocity and pressure distributions can be found between different globe valves. The Reynolds number is around 7×10^5, which means the flow regime is turbulent. At the bottom of the valve body, there is a high-pressure and low-velocity zone, which results from the sudden change of the flow direction. When there is no cage, most water flows directly toward the valve downstream, resulting in a small high-pressure zone at the bottom of the valve core closing to the valve outlet. While cages are added, the valve resistance increases, and the pressure distribution at the valve core bottom becomes more even. From Figure 4, it can also be found that the cage number has little influence on the streamline distribution once the cage is added.

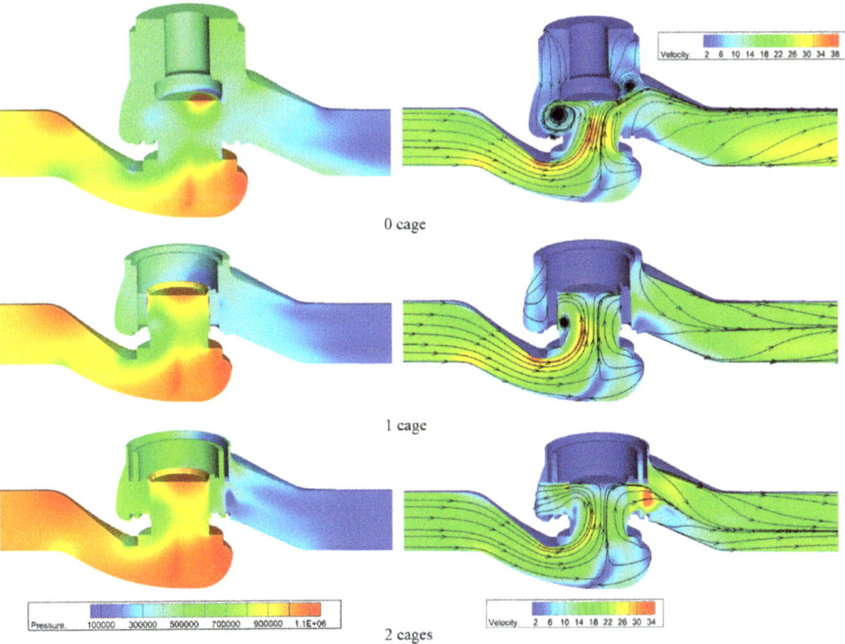

Figure 4. Velocity and pressure distributions for the globe valve with a different number of cages.

When water vapor volume fraction is above 0.1, the water vapor volume distribution in different globe valves is compared qualitatively [4,37] and shown in Figure 5. It is clear that cavitation caused water vapor mainly congregates around the valve seat and the bottom of the valve core if there is no cage, but the water vapor only appears at the location near the valve outlet, which is because of the high velocity and low pressure as shown in Figure 4. Besides, a relatively large water vapor volume can be found at the location where the valve body and the downstream pipe connect. When a cage is added, the region where water vapor appears varies and is limited close to the cage itself, which protects the valve body from being damaged. When two cages are added, the region where cavitation appears reduces further but is still near the two cages.

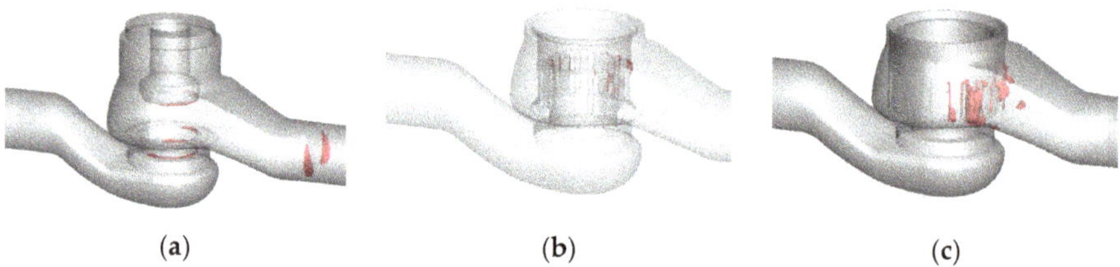

(a) (b) (c)

Figure 5. Water vapor volume distribution for globe valve with a different number of cages: (**a**) no cage; (**b**) 1 cage; (**c**) 2 cages.

To compare cavitation intensity in a globe valve with a different number of cages, dimensionless cavitation number σ is used. The expression of the cavitation number is shown below.

$$\sigma = \frac{p_o - p_v}{1/2\rho v^2} \tag{4}$$

here p_o is the pressure at valve outlet, p_v is the saturated vapor pressure, ρ is the density of water, v is the velocity at the valve outlet.

The loss coefficient ξ is also calculated to analyze the flow characteristics of different globe valves. The expression of the loss coefficient is shown below.

$$\xi = \frac{p_i - p_o}{1/2\rho v^2} \tag{5}$$

here p_i is the pressure at valve inlet.

The comparison of the cavitation number and the loss coefficient between different globe valves is shown in Figure 6. It can be found that the cavitation number and the loss coefficient increase with the increase in the valve cage number, which means that cavitation intensity can be suppressed by adding a valve cage in a globe valve. In Figure 5, it is noticed that the water vapor volume in the globe valve with two cages is higher than other globe valves. The reason is that when the total pressure in the valve inlet is the same, the velocity inside the globe valve with two cages is lower due to high resistance, which leads to a higher static pressure at valve inlet. Thus, the actual water vapor volume is higher in the globe valve with two cages. Together with Figures 5 and 6, it can also be found that the cavitation number is useful when deciding the cavitation intensity in valves.

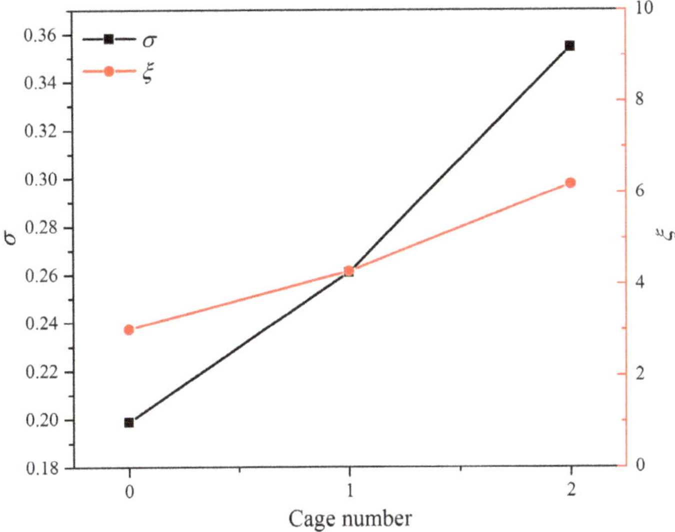

Figure 6. Comparison of cavitation number and loss coefficient for the globe valve with a different number of cages.

3.2. *Effects of the Working Temperature and Outlet Pressure*

When the outlet pressure and the working temperature vary, the total pressure drop or the saturated vapor pressure changes, thus leading to the variation of cavitation intensity. In this section, the effects of the working temperature and the outlet pressure are studied and compared.

Figures 7 and 8 show the water vapor volume distribution in the two-cage globe valve under different outlet pressures and different working temperatures. The working temperature for Figure 7 is 80 °C, and the outlet pressure for Figure 8 is 0.1 MPa. The water vapor volume decreases with the increase in the outlet pressure or the decrease in the working temperature. Based on the water vapor volume distributions, one can find that increased outlet pressure has more positive effects on cavitation intensity than the working temperature.

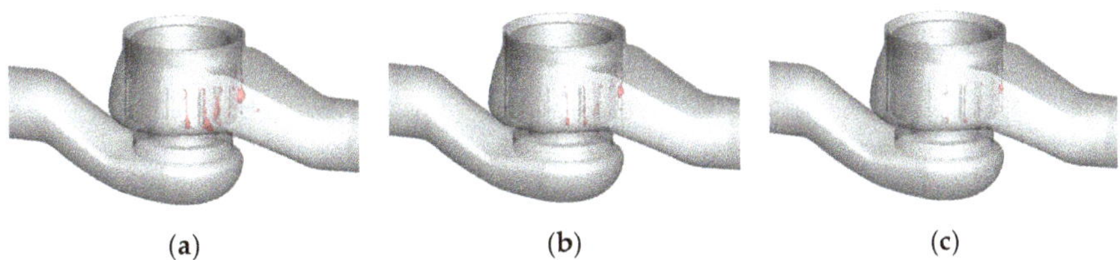

Figure 7. Water vapor distribution in two-cage globe valve with different outlet pressure: (**a**) 0.15 MPa; (**b**) 0.20 MPa; (**c**) 0.25 MPa.

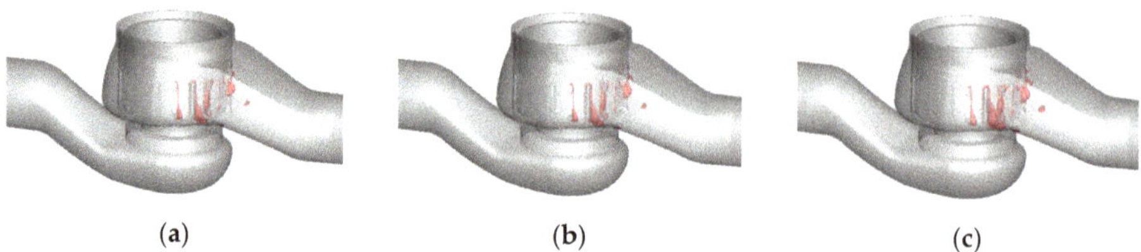

Figure 8. Water vapor volume distribution in two-cage globe valve with different working temperature: (**a**) 20 °C; (**b**) 40 °C; (**c**) 60 °C.

To investigate the effects of the outlet pressure and the working temperature quantitatively, the cavitation number and the loss coefficient are compared in Figure 9, where the straight line stands for cavitation number, and the dash line stands for the loss coefficient. It can be found that the cavitation number increases with the increase in the outlet pressure and the decrease in the working temperature, which is consistent with the above results. Cavitation number almost has a linear relationship with the outlet pressure. When the outlet pressure increases by 0.05 MPa, the cavitation number increases nearly by 0.4. While the increment of the cavitation number decreases with the decrease in the working temperature. When the working temperature decreases from 80 °C to 60 °C, the average increment of cavitation number is 0.134, and the average increment of cavitation number is 0.0538 if the working temperature decreases from 40 °C to 20 °C.

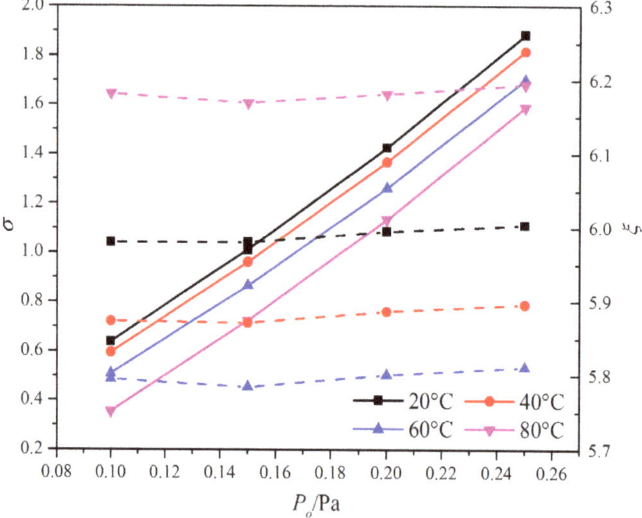

Figure 9. Cavitation number and loss coefficient of two-cage globe valve with different outlet pressure and working temperature.

The loss coefficient barely changes with varying outlet pressure which is because the loss coefficient is affected by the pressure difference between the valve inlet and valve outlet shown in Equation (5). However, with the increase in the working temperature, the loss coefficient decreases firstly then increases rapidly, and when the working temperature is 80 °C, the loss coefficient has the maximum value. The difference of the loss coefficient may be caused by the cavitation phenomenon inside valves.

In the numerical simulation applied in this study, the finite volume method is used. Thus, the water vapor volume in valves can be calculated based on the water vapor fraction and the discrete volume, and the expression is shown below.

$$V = \sum_{1}^{n} \alpha_v dV \tag{6}$$

here V stands for the total water vapor volume in a valve, n stands for the total number of the discrete volume, dV stands for the volume of a discrete volume, α_v stands for the water vapor fraction in a discrete volume. The water vapor volume in the investigated two-cage globe valve under different outlet pressures and the different working temperatures is shown in Figure 10. It can be found that when the outlet pressure is larger than 0.2 MPa, the variation of water vapor volume under different working temperatures is small. While the outlet pressure is smaller than 0.2 MPa, the water vapor volume increases with the increase in the working temperature and the outlet pressure, and the higher the working temperature, the greater the influence of the outlet pressure on water vapor volume. Together with Figure 8, it can be concluded that increase in the outlet pressure is more effective in suppressing the cavitation than the decrease in the working temperature.

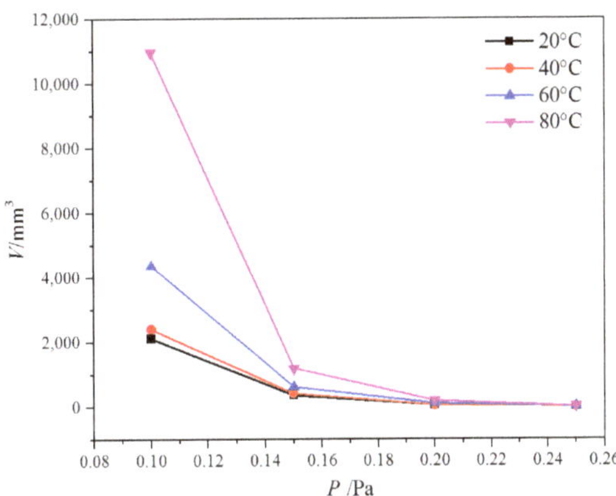

Figure 10. Water vapor volume in the two-cage globe valve with different outlet pressure and working temperature.

3.3. Effects of the Installation Angle

When there are two cages inside a globe valve, the installation angle between the two cages influences the flow resistance, thus affecting the cavitation intensity. For the investigated two-cage globe valve, the number of the throttling window on each cage is 20, thus the installation angle from 0° to 9° is investigated. Figure 11 shows the streamline on a cross-section in the two-cage globe valve under different installation angles. It can be seen that when the installation is 0°, most water flows directly from the inner cage to the outer cage, resulting in small flow resistance. With the increase in the installation angle, water comes from the inner cage and collides with the outer cage, and then changes flow direction two times before finally flowing to the valve downstream, resulting in larger flow resistance.

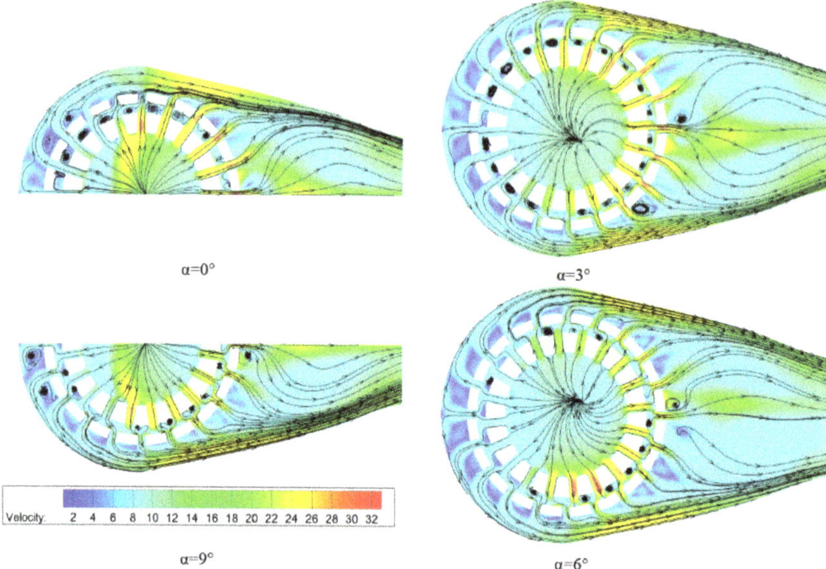

Figure 11. Streamline on a cross-section for the two-cage globe valve with different installation angles.

Figure 12 shows the loss coefficient in two-cage globe valves under different installation angles. It can be found that under different installation angles, the effects of the working temperature on the loss coefficient are the same with the above analysis, and the loss coefficient has the maximum value when the working temperature is 80 °C. Besides, with the increase in the installation angle, the loss coefficient becomes dependent on the outlet pressure even under the same working temperature, and the larger the outlet pressure, the smaller the loss coefficient. Together with Figures 11 and 12, it can also be concluded that the larger the installation angle, the larger the flow resistance, which means the smaller the cavitation intensity.

Cavitation number in two-cage globe valves with different installation angles is shown in Figure 13, it can be found that with the increase in the outlet pressure and the decrease in the working temperature, cavitation number increases, which is consistent with the above results. With the increase in the installation angle, the cavitation number also increases, which means cavitation intensity in two-cage globe valves with a large installation angle is small. The increment of cavitation number decreases with the decrease in the outlet pressure but barely changes with the variation of the working temperature. When the outlet pressure is 0.25 MPa, the increment of cavitation number is about 0.39 (the installation angle varies from 0° to 9°), but the increment of cavitation number is only 0.144 when the outlet pressure is 0.1 MPa. Based on the results analyzed in Section 3.2, the most effective method to suppress cavitation intensity in the two-cage globe valve is by increasing the outlet pressure.

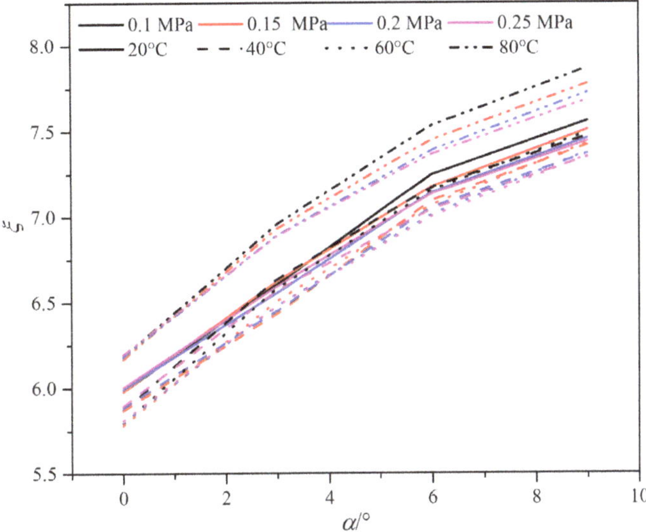

Figure 12. Loss coefficient in the two-cage globe valve with different installation angles.

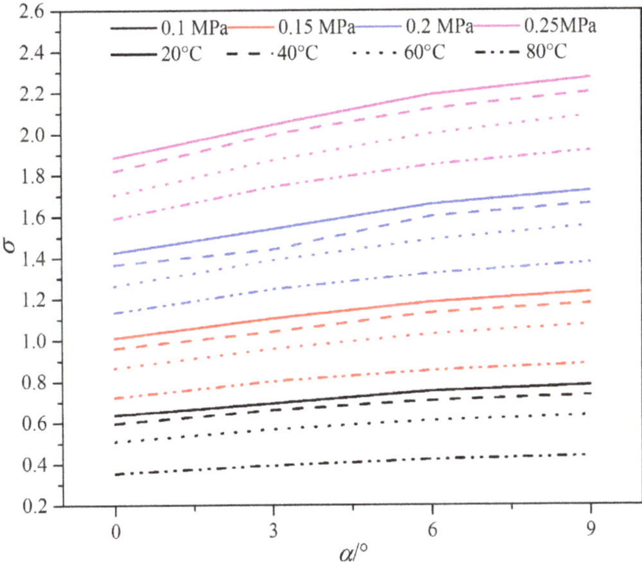

Figure 13. Cavitation number in the two-cage globe valve with different installation angles.

4. Conclusions

Cavitation phenomenon and cavitation intensity in a two-cage globe valve are investigated in this study using the computational fluid dynamics method. A comparison of the water vapor volume distribution caused by cavitation for a globe valve with a different number of cages is made, and cavitation can be restricted around the valve cage and the valve body damage can be suppressed by adding a cage in the valve body, and cavitation intensity decreases with the increase in valve cage number. The effects of the valve cage with 20 slender orifices, the outlet pressure, and the working temperature on cavitation intensity and loss coefficient inside the investigated two-cage globe valve are studied, and cavitation number and water vapor volume are chosen to stand for the cavitation intensity. With the increasing outlet pressure and the installation angle, or the decreasing working temperature, cavitation number increases which means cavitation intensity decreases. When cavitation occurs, the loss coefficient firstly decreases with the increase in the working temperature, and then rapidly increases, while the loss coefficient barely changes with varying outlet pressure but increases with the increasing installation angle. Besides, among three focused parameters, reducing the outlet pressure can remarkably decrease the cavitation intensity compared to the others and the working temperature has minimal effects on cavitation intensity.

Author Contributions: Conceptualization, J.-y.L. and Z.-x.G.; methodology, J.-y.L.; software, Z.-x.G.; validation, J.-y.L.; formal analysis, Z.-x.G.; resources, H.W. and Z.-j.J.; writing—original draft preparation, J.-y.L. and Z.-x.G.; writing—review and editing, Z.-x.G.; supervision, J.-y.L. All authors have read and agreed to the published version of the manuscript.

Funding: This research received no external funding.

Conflicts of Interest: The authors declare no conflict of interest. The funders had no role in the design of the study; in the collection, analyses, or interpretation of data; in the writing of the manuscript, or in the decision to publish the results.

References

1. Qian, J.; Gao, Z.; Hou, C.; Jin, Z. A comprehensive review of cavitation in valves: Mechanical heart valves and control valves. *Bio-Des. Manuf.* **2019**, *2*, 119–136. [CrossRef]
2. Cui, B.; Lin, Z.; Zhu, Z.; Wang, H.; Ma, G. Influence of opening and closing process of ball valve on external performance and internal flow characteristics. *Exp. Therm. Fluid Sci.* **2017**, *80*, 193–202. [CrossRef]
3. Qian, J.; Wu, J.; Gao, Z.; Jin, Z. Effects of throttling window on flow rate through feed-water valves. *ISA Trans.* **2020**. [CrossRef]
4. Yaghoubi, H.; Madani, S.A.H.; Alizadeh, M. Numerical study on cavitation in a globe control valve with different numbers of anti-cavitation trims. *J. Cent. South Univ.* **2018**, *25*, 2677–2687. [CrossRef]
5. Qian, J.; Chen, M.; Jin, Z.; Chen, L.; Sundén, B. A numerical study of heat transfer effects and aerodynamic noise reduction in superheated steam flow passing a temperature and pressure regulation valve. *Numer. Heat Transf. Part A* **2020**, *77*, 873–889. [CrossRef]
6. Fratini, L.; Pasta, S. Residual stresses in friction stir welded parts of complex geometry. *Int. J. Adv. Manuf. Technol.* **2012**, *59*, 547–557. [CrossRef]
7. Wang, X.; Li, W.; Li, C.; Peng, Y. Structure optimization and flow field simulation of plate type high speed on-off valve. *J. Cent. South Univ.* **2020**, *27*, 1557–1571. [CrossRef]
8. Lin, Z.; Sun, X.; Yu, T.; Zhang, Y.; Li, Y.; Zhu, Z. Gas–solid two-phase flow and erosion calculation of gate valve based on the CFD-DEM model. *Powder Technol.* **2020**, *366*, 395–407. [CrossRef]
9. Jin, Z.; Gao, Z.; Chen, M.; Qian, J. Parametric study on Tesla valve with reverse flow for hydrogen decompression. *Int. J. Hydrog. Energy* **2018**, *43*, 8888–8896. [CrossRef]
10. Qian, J.; Chen, M.; Liu, X.; Jin, Z. A numerical investigation of the flow of nanofluids through a micro Tesla valve. *J. Zhejiang Univ. Sci. A* **2019**, *20*, 50–60. [CrossRef]
11. Qian, J.; Wu, J.; Gao, Z.; Wu, A.; Jin, Z. Hydrogen decompression analysis by multi-stage Tesla valves for hydrogen fuel cell. *Int. J. Hydrog. Energy* **2019**, *44*, 13666–13674. [CrossRef]
12. Qian, J.; Chen, M.; Gao, Z.; Jin, Z. Mach number and energy loss analysis inside multi-stage Tesla valves for hydrogen decompression. *Energy* **2019**, *179*, 647–654. [CrossRef]
13. Fan, S.; Xu, R.; Ji, H.; Yang, S.; Yuan, Q. Experimental Investigation on Contaminated Friction of Hydraulic Spool Valve. *Appl. Sci.* **2019**, *9*, 5230. [CrossRef]
14. Yan, H.; Li, J.; Cai, C.; Ren, Y. Numerical Investigation of Erosion Wear in the Hydraulic Amplifier of the Deflector Jet Servo Valve. *Appl. Sci.* **2020**, *10*, 1299. [CrossRef]
15. Pasta, S.; Gentile, G.; Raffa, G.M.; Bellavia, D.; Chiarello, G.; Liotta, R.; Luca, A.; Scardulla, C.; Pilato, M. In silico shear and intramural stresses are linked to aortic valve morphology in dilated ascending aorta. *Eur. J. Vasc. Endovasc. Surg.* **2017**, *54*, 254–263. [CrossRef]

16. Rinaudo, A.; D'Ancona, G.; Lee, J.J.; Pilato, G.; Amaducci, A.; Baglini, R.; Follis, F.; Pilato, M.; Pasta, S. Predicting outcome of aortic dissection with patent false lumen by computational flow analysis. *Cardiovasc. Eng. Technol.* **2014**, *5*, 176–188. [CrossRef]
17. Zhang, J.; Wang, D.; Xu, B.; Gan, M.; Pan, M.; Yang, H. Experimental and numerical investigation of flow forces in a seat valve using a damping sleeve with orifices. *J. Zhejiang Univ. Sci. A* **2018**, *19*, 417–430. [CrossRef]
18. Zhong, Q.; Zhang, B.; Bao, H.; Hong, H.; Ma, J.; Ren, Y.; Yang, H.; Fung, R. Analysis of pressure and flow compound control characteristics of an independent metering hydraulic system based on a two-level fuzzy controller. *J. Zhejiang Univ. Sci. A* **2019**, *20*, 184–200. [CrossRef]
19. Saha, B.K.; Peng, J.; Li, S. Numerical and experimental investigations of cavitation phenomena inside the pilot stage of the deflector jet servo-valve. *IEEE Access* **2020**, *8*, 64238–64249. [CrossRef]
20. Kim, H.; Kim, S. Optimization of pressure relief valve for pipeline system under transient induced cavitation condition. *Urban Water J.* **2019**, *16*, 718–726. [CrossRef]
21. Yi, J.; Hu, H.; Zheng, Y.; Zhang, Y. Experimental and computational failure analysis of a high pressure regulating valve in a chemical plant. *Eng. Fail. Anal.* **2016**, *70*, 188–199. [CrossRef]
22. Zhang, X.; Zhang, T.; Hou, Y.; Zhu, K.; Huang, Z.; Wu, K. Local loss model of dividing flow in a bifurcate tunnel with a small angle. *J. Zhejiang Univ. Sci. A* **2019**, *20*, 21–35. [CrossRef]
23. Liu, G.; Lin, Y.; Guan, G.; Yu, Y. Numerical research on the anti-sloshing effect of a ring baffle in an independent type C LNG tank. *J. Zhejiang Univ. Sci.* **2018**, *19*, 758–773. [CrossRef]
24. Wen, H.; Ma, J.; Jiang, G.; Lee, Y. Effect of needle valve motion on cavitation and gas ingestion at end of injection. *Combust. Sci. Technol.* **2020**. [CrossRef]
25. Li, W.; Gao, Z.; Jin, Z.; Qian, J. Transient study of flow and cavitation inside a bileaflet mechanical heart valve. *Appl. Sci.* **2020**, *10*, 2548. [CrossRef]
26. Jin, Z.; Gao, Z.; Qian, J.; Wu, Z.; Sunden, B. A parametric study of hydrodynamic cavitation inside globe valves. *ASME J. Fluids Eng.* **2018**, *140*, 031208. [CrossRef]
27. Jin, Z.; Qiu, C.; Jiang, C.; Wu, J.; Qian, J. Effect of valve core shapes on cavitation flow through a sleeve regulating valve. *J. Zhejiang Univ. Sci. A* **2020**, *21*, 1–14. [CrossRef]
28. Liu, X.; Wu, Z.; Li, B.; Zhao, J.; He, J.; Li, W.; Zhang, C.; Xie, F. Influence of inlet pressure on cavitation characteristics in regulating valve. *Eng. Appl. Comput. Fluid Mech.* **2020**, *14*, 299–310. [CrossRef]
29. Qiu, C.; Jiang, C.; Zhang, H.; Wu, J.; Jin, Z. Pressure drop and cavitation analysis on sleeve regulating valve. *Processes* **2019**, *7*, 829. [CrossRef]
30. Liu, B.; Zhao, J.; Qian, J. Numerical analysis of cavitation erosion and particle erosion in butterfly valve. *Eng. Fail. Anal.* **2017**, *80*, 312–324. [CrossRef]
31. Jin, Z.; Gao, Z.; Li, X.; Qian, J. Cavitating Flow through a Micro-Orifice. *Micromachines* **2019**, *10*, 191. [CrossRef] [PubMed]
32. Zhang, J. Flow characteristics of a hydraulic cone-throttle valve during cavitation. *Ind. Lubr. Tribol.* **2019**, *71*, 1186–1193. [CrossRef]
33. Yuan, C.; Song, J.; Liu, M. Comparison of compressible and incompressible numerical methods in simulation of a cavitating jet through a poppet valve. *Eng. Appl. Comput. Fluid Mech.* **2019**, *13*, 67–90. [CrossRef]
34. Xu, H.; Wang, H.; Hu, M.; Jiao, L.; Li, C. Optimal design and experimental research of the anti-cavitation structure in the water hydraulic relief valve. *J. Pressure Vessel Technol.* **2018**, *140*, 051601. [CrossRef]
35. Weijie, S.; Shuping, C.; Xiaohui, L.; Zuti, Z.; Yuquan, Z. Experimental research on the cavitation suppression in the water hydraulic throttle valve. *J. Pressure Vessel Technol.* **2017**, *139*, 051302. [CrossRef]
36. Lee, M.G.; Lim, C.S.; Han, S.H. Shape design of the bottom plug used in a 3-way reversing valve to minimize the cavitation effect. *Int. J. Precis. Eng. Manuf.* **2016**, *17*, 401–406. [CrossRef]
37. Chern, M.; Hsu, P.; Cheng, Y.; Tseng, P.; Hu, C. Numerical Study on Cavitation Occurrence in Globe Valve. *J. Energy Eng.* **2013**, *139*, 25–34. [CrossRef]
38. Zwart, P.J.; Gerber, A.G.; Belamri, T. A Two-Phase Flow Model for Predicting Cavitation Dynamics. In Proceedings of the ICMF International Conference on Multiphase Flow, Firenze, Italy, 22–27 May 2004; p. 152.

MDPI
St. Alban-Anlage 66
4052 Basel
Switzerland
Tel. +41 61 683 77 34
Fax +41 61 302 89 18
www.mdpi.com

Applied Sciences Editorial Office
E-mail: applsci@mdpi.com
www.mdpi.com/journal/applsci

www.ingramcontent.com/pod-product-compliance
Lightning Source LLC
LaVergne TN
LVHW070843170726
843515LV00005B/558

* 9 7 8 3 0 3 6 5 1 5 4 6 5 *